电灯

交流发电机

PATENTS BRIGHTENED OUR LIVES

影响世界的专利

国家知识产权局专利局专利文献部 ◎ 编著

吸尘器

机械电视

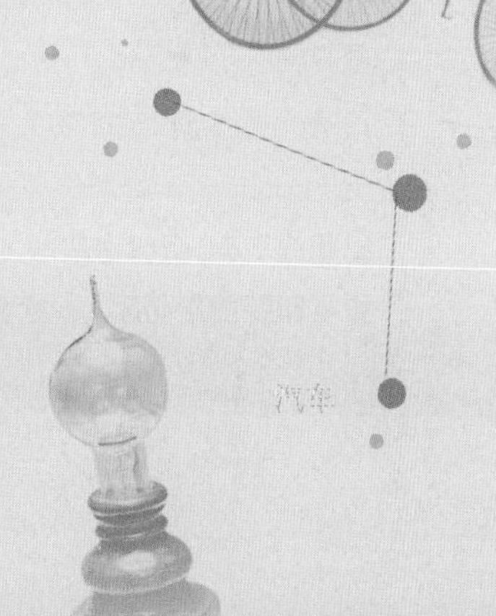

知识产权出版社
全国百佳图书出版单位

图书在版编目（CIP）数据

影响世界的专利 / 国家知识产权局专利局专利文献部编著 . —北京：知识产权出版社，2015.11

ISBN 978-7-5130-3873-7

Ⅰ. ①影… Ⅱ. ①国… Ⅲ. ①专利制度—介绍—世界 Ⅳ. ① G306.3

中国版本图书馆 CIP 数据核字（2015）第 253393 号

责任编辑：牛洁颖　　**责任校对**：韩秀天

装帧设计：品　序　　**责任出版**：刘译文

影响世界的专利

国家知识产权局专利局专利文献部　编著

出版发行：知识产权出版社有限责任公司　网　址：http://www.ipph.cn

社　址：北京市海淀区马甸南村1号（邮编:100088）　天猫旗舰店：https://zscqcbs.tmall.com

责编电话：010-82000860-8109　责编邮箱：niujieying@sina.com

发行电话：010-82000860转8101/8102　发行传真：010-82000893/82005070/82000270

印　刷：天津市银博印刷集团有限公司　经　销：各大网上书店、新华书店及相关专业书店

开　本：787mm×1092mm　1/16　印　张：12

版　次：2015年11月第1版　印　次：2016年11月第2次印刷

字　数：220千字　定　价：36.00元

ISBN 978-7-5130-3873-7

编 委 会

目录

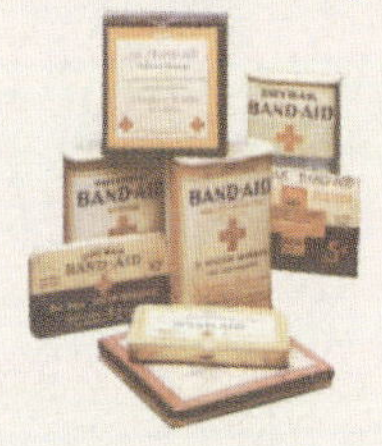
BAND-AID
BAND-AID
BAND-AID
BAND-AID

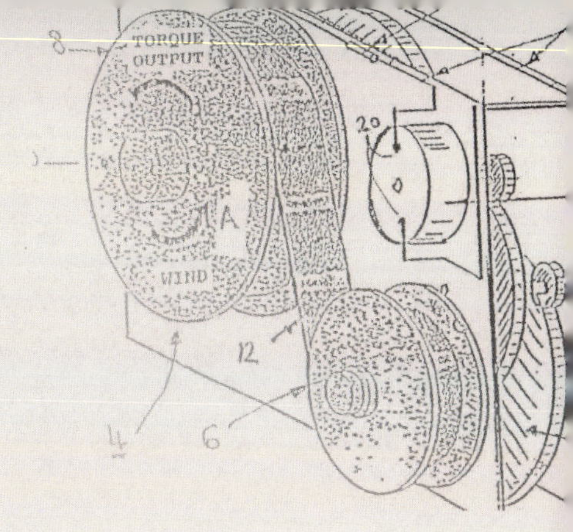
TORQUE
OUTPUT
WIND
A
8
20
12
4
6

电灯的发明

Electric lamp invention

伟大发明家爱迪生

Thomas Alva Edison

1847 年 2 月 11 日，出生于美国俄亥俄州的米兰镇。他一生只在学校里念过三个月的书，但他勤奋好学，勤于思考，一生为人类创造了 1000 多件发明。

1931 年 10 月 18 日，世界著名发明家托马斯·阿尔瓦·爱迪生在美国的西奥林奇去世，享年 84 岁。有人作过统计，爱迪生一生的发明，在专利局正式登记的有 1300 种左右。爱迪生拥有白炽灯、留声机、炭粒话筒、电影放映机等 1093 种发明专利权。

1881 年是爱迪生发明的最高纪录年，这一年，他申请立案的发明就有 141 种，平均每三天就有一种新发明问世。他的一生是光荣的，他的一切是为人类的。

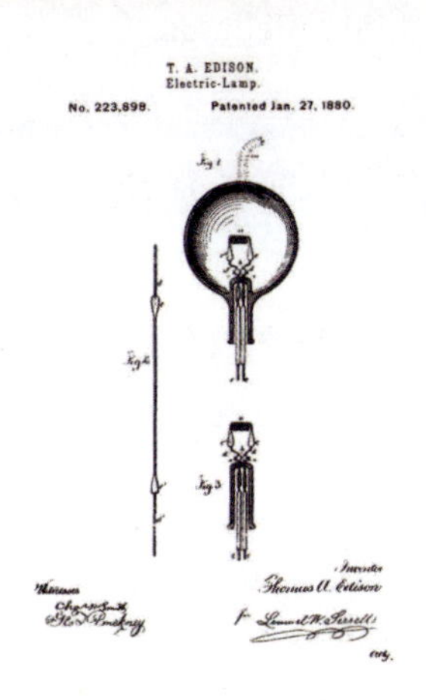

T. A. EDISON.
Electric-Lamp.
No. 223,898. Patented Jan. 27, 1880.

US 223898

www.sipo.gov.cn

第一篇 电 灯

电灯是人类最伟大的发明之一。在电灯问世以前，人们普遍使用煤油灯和煤气灯作为照明工具。燃料在燃烧过程中会产生浓烈的黑烟和刺鼻的气味，并且使用起来也很不方便，还容易引发火灾。人们迫切需要一种既安全又方便的照明灯。

1812 年，英国化学家汉弗莱·戴维（Humphry Davy）将 2000 节电池和两根碳棒连接起来，制成了世界上第一盏电光源——弧光灯。但弧光灯的光线太强，无法用于室内照明，只能在街道或广场等地方使用。1845 年，美国发明家约翰·威灵顿·斯塔尔（John Wellington Starr）在真空泡内使用碳丝制成了电灯，但由于当时抽真空的技术很差，灯泡中残余的空气使得灯丝很容易烧断，这种灯的寿命仅有个把小时，不具有实用价值。1878 年英国科学家约瑟夫·斯旺（Joseph Swan）试制成功了第一只白炽灯并获得英国 GB4933 号专利权。无数科学家绞尽脑汁，希望能制造一种物美价廉、经久耐用的家用电灯，最终爱迪生实现了这个愿望。

托马斯·阿尔瓦·爱迪生（Thomas Alva Edison，1847–1931 年）于 1879 年开始投入对电灯的研究。他在认真总结了前人经验之后认为，延长白炽灯寿命的关键是提高灯泡的真空度和采用耗电少、发光强且价格便宜的耐热材料做灯丝。爱迪生先后试用了 1600 多种耐热材料，结果都不理想，最终他将目光放在了棉纱上。1879 年 10 月的一天，

他用炭化棉线做灯丝放入玻璃球内，并将球内抽成真空，结果这种灯丝发出了明亮而稳定的光，竟能连续使用45个小时。由此，世界上第一批炭丝白炽灯诞生了。1879年爱迪生在美国申请了专利，并于1880年获得名为“电灯”的美国US223898号专利权，如图1-1、图1-2所示。

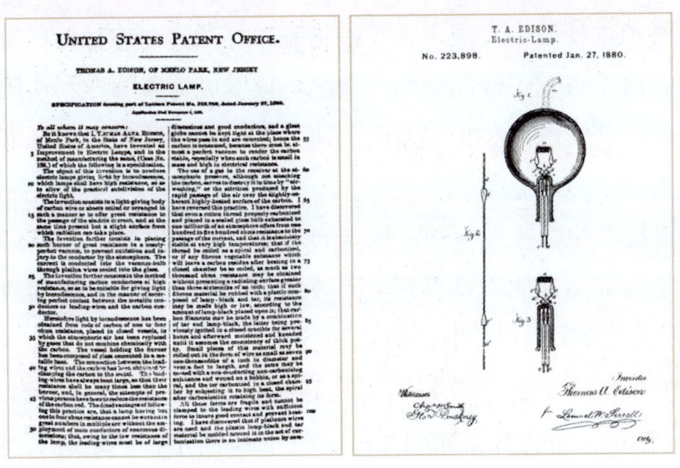

UNITED STATES PATENT OFFICE.

THOMAS A. EDISON, OF MENLO PARK, NEW JERSEY.

ELECTRIC LAMP.

图1-1　　图1-2

从这篇专利说明书中我们可以看出，爱迪生发明的白炽灯的基本结构是由高电阻材料（炭丝）的灯丝和低电阻材料的电源线和导线组成，当电流流过灯丝时，由于材料电阻率高，产生大量热量，从而发光。

爱迪生并没有止步于此，为了使电灯亮的时间更长，他决定寻找更合适的灯丝材料。根据棉线的性质，爱迪生决定从植物纤维这方面去寻找新的材料。经过大量的试验，最后爱迪生选择了竹纤维。他把炭化后的竹丝装进玻璃泡，通电后，这种竹丝灯泡竟连续不断地亮了1200个小时。经过进一步考察和比较，爱迪生发现在日本生长的一种竹子最适合做灯丝，于是从日本大量进口这种竹子，并开设电厂，架设电线。很快，美国人民便用上了这种价廉物美、经久耐用的竹丝灯泡。在竹丝灯使用了多年之后，1906年，爱迪生又改用钨丝来做灯丝，使灯泡的质量再次得到提高，并一直沿用到今天。

白炽灯的发明带来了巨大的市场价值，它的专利权无疑是生产者们占领市场的一大法宝，由此引发了一系列关于专利权的争抢诉讼。

由于斯旺和爱迪生发明电灯的时间相差不多，两位发明家曾在英国展开了专利侵权诉讼，最终斯旺胜诉，爱迪生在英国的电灯公司因此被迫让斯旺加入作为合伙人。而在美国，爱迪生的专利权也受到挑战，1895 年索耶—曼公司（Sawyer & Man）控告爱迪生电灯公司侵犯其专利权，最终爱迪生胜诉。正是由于爱迪生通过坚持不懈的试验探索，在数千种植物纤维中发现了竹纤维是制作灯丝的理想材料，美国法院最终在判决中肯定了他的发明与贡献。该案成为美国专利法历史上的经典判例，其形成的关于“发明创造是否可实现”的判定标准在美国专利法实践中沿用至今。

随着技术的成熟，钨丝白炽灯的改进空间越来越小，已经很难进一步大幅降低生产成本、提高照明效果。以荧光灯为代表的放电灯开始逐步发展起来，一些工程师也转而对放电灯展开研究。其实早在 20 世纪初，爱迪生就已经开始了对荧光灯的研究，并于 1907 年获得美国 US865367 号专利权，如图 1–3、图 1–4 所示，只是迫于当时技术的限制，爱迪生最终放弃了此项发明。

UNITED STATES PATENT OFFICE.

图 1–3

图 1–4

但是人们并没有中断对荧光灯的研究。通用电气、飞利浦、西屋电气等制造商都分别研制出了不同类型的荧光灯，到 20 世纪 50 年代，美国的荧光灯产量已经超过了白炽灯。

如今，人类的生存环境面临诸多挑战，推行低碳经济已经成为全

球共识。在照明领域，白炽灯由于能量转化率低已面临着被淘汰的境遇。进入 21 世纪后，节能灯、发光二极管（LED 灯）、高强度气体放电灯等照明方式开始发展流行。它们以各自节能、高效、寿命长等优点，逐步取代了白炽灯在照明市场的统治地位，但是白炽灯的离去并不意味着电光文明时代的结束，相反，随着科技的发展，相信会有越来越多的新型照明设备走进人类的生活。

美国 America

1880. US 223898

——电灯

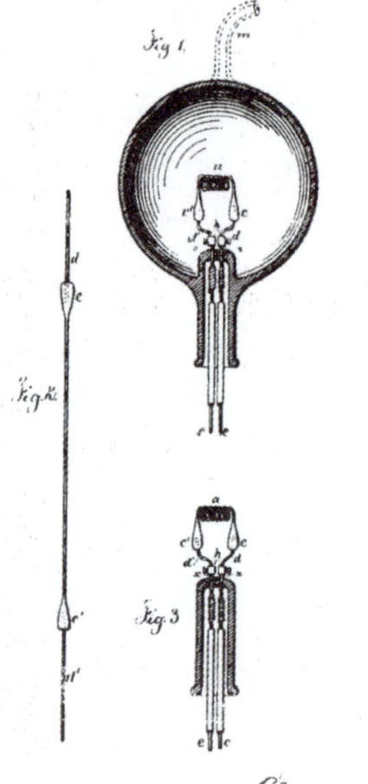
T. A. EDISON.
Electric-Lamp.
No. 223,898. Patented Jan. 27, 1880.

UNITED STATES PATENT OFFICE.

THOMAS A. EDISON, OF MENLO PARK, NEW JERSEY

ELECTRIC LAMP.

SPECIFICATION forming part of Letters Patent No. 223,898, dated January 27, 1880.

Application filed November 4, 1879.

To all whom it may concern:

Be it known that I, THOMAS ALVA EDISON, of Menlo Park, in the State of New Jersey, United States of America, have invented an Improvement in Electric Lamps, and in the method of manufacturing the same, (Case No. 186,) of which the following is a specification.

The object of this invention is to produce electric lamps giving light by incandescence, which lamps shall have high resistance, so as to allow of the practical subdivision of the electric light.

The invention consists in a light-giving body of carbon wire or sheets coiled or arranged in such a manner as to offer great resistance to the passage of the electric current, and at the same time present but a slight surface from which radiation can take place.

The invention further consists in placing such burner of great resistance in a nearly-perfect vacuum, to prevent oxidation and injury to the conductor by the atmosphere. The current is conducted into the vacuum-bulb through platina wires sealed into the glass.

The invention further consists in the method of manufacturing carbon conductors of high resistance, so as to be suitable for giving light by incandescence, and in the manner of securing perfect contact between the metallic conductors or leading-wires and the carbon conductor.

Heretofore light by incandescence has been obtained from rods of carbon of one to four ohms resistance, placed in closed vessels, in which the atmospheric air has been replaced by gases that do not combine chemically with the carbon. The vessel holding the burner has been composed of glass cemented to a metallic base. The connection between the leading-wires and the carbon has been obtained by clamping the carbon to the metal. The leading-wires have always been large, so that their resistance shall be many times less than the burner, and, in general, the attempts of previous persons have been to reduce the resistance of the carbon rod. The disadvantages of following this practice are, that a lamp having but one to four ohms resistance cannot be worked in great numbers in multiple arc without the employment of main conductors of enormous dimensions; that, owing to the low resistance of the lamp, the leading-wires must be of large dimensions and good conductors, and a glass globe cannot be kept tight at the place where the wires pass in and are cemented; hence the carbon is consumed, because there must be almost a perfect vacuum to render the carbon stable, especially when such carbon is small in mass and high in electrical resistance.

The use of a gas in the receiver at the atmospheric pressure, although not attacking the carbon, serves to destroy it in time by "air-washing," or the attrition produced by the rapid passage of the air over the slightly-coherent highly-heated surface of the carbon. I have reversed this practice. I have discovered that even a cotton thread properly carbonized and placed in a sealed glass bulb exhausted to one-millionth of an atmosphere offers from one hundred to five hundred ohms resistance to the passage of the current, and that it is absolutely stable at very high temperatures; that if the thread be coiled as a spiral and carbonized, or if any fibrous vegetable substance which will leave a carbon residue after heating in a closed chamber be so coiled, as much as two thousand ohms resistance may be obtained without presenting a radiating-surface greater than three-sixteenths of an inch; that if such fibrous material be rubbed with a plastic composed of lamp-black and tar, its resistance may be made high or low, according to the amount of lamp-black placed upon it; that carbon filaments may be made by a combination of tar and lamp-black, the latter being previously ignited in a closed crucible for several hours and afterward moistened and kneaded until it assumes the consistency of thick putty. Small pieces of this material may be rolled out in the form of wire as small as seven one-thousandths of a inch in diameter and over a foot in length, and the same may be coated with a non-conducting non-carbonizing substance and wound on a bobbin, or as a spiral, and the tar carbonized in a closed chamber by subjecting it to high heat, the spiral after carbonization retaining its form.

All these forms are fragile and cannot be clamped to the leading wires with sufficient force to insure good contact and prevent heating. I have discovered that if platinum wires are used and the plastic lamp-black and tar material be molded around it in the act of carbonization there is an intimate union by combination and by pressure between the carbon and platina, and nearly perfect contact is obtained without the necessity of clamps; hence the burner and the leading-wires are connected to the carbon ready to be placed in the vacuum-bulb.

When fibrous material is used the plastic lamp-black and tar are used to secure it to the platina before carbonizing.

By using the carbon wire of such high resistance I am enabled to use fine platinum wires for leading-wires, as they will have a small resistance compared to the burner, and hence will not heat and crack the sealed vacuum-bulb. Platina can only be used, as its expansion is nearly the same as that of glass.

By using a considerable length of carbon wire and coiling it the exterior, which is only a small portion of its entire surface, will form the principal radiating-surface; hence I am able to raise the specific heat of the whole of the carbon, and thus prevent the rapid reception and disappearance of the light, which on a plain wire is prejudicial, as it shows the least unsteadiness of the current by the flickering of the light; but if the current is steady the defect does not show.

I have carbonized and used cotton and linen thread, wood splints, papers coiled in various ways, also lamp-black, plumbago, and carbon in various forms, mixed with tar and kneaded so that the same may be rolled out into wires of various lengths and diameters. Each wire, however, is to be uniform in size throughout.

If the carbon thread is liable to be distorted during carbonization it is to be coiled between a helix of copper wire. The ends of the carbon or filament are secured to the platina leading-wires by plastic carbonizable material, and the whole placed in the carbonizing-chamber. The copper, which has served to prevent distortion of the carbon thread, is afterward eaten away by nitric acid, and the spiral soaked in water, and then dried and placed on the glass holder, and a glass bulb blown over the whole, with a leading-tube for exhaustion by a mercury-pump. This tube, when a high vacuum has been reached, is hermetically sealed.

With substances which are not greatly distorted in carbonizing, they may be coated with a non-conducting non-carbonizable substance, which allows one coil or turn of the carbon to rest upon and be supported by the other.

In the drawings, Figure 1 shows the lamp sectionally. *a* is the carbon spiral or thread. *c c'* are the thickened ends of the spiral, formed of the plastic compound of lamp-black and tar. *d d'* are the platina wires. *h h* are the clamps, which serve to connect the platina wires, cemented in the carbon, with the leading-wires *x x*, sealed in the glass vacuum-bulb. *e e* are copper wires, connected just outside the bulb to the wires *x x*. *m* is the tube (shown by dotted lines) leading to the vacuum-pump, which, after exhaustion, is hermetically sealed and the surplus removed.

Fig. 2 represents the plastic material before being wound into a spiral.

Fig. 3 shows the spiral after carbonization, ready to have a bulb blown over it.

I claim as my invention—

1. An electric lamp for giving light by incandescence, consisting of a filament of carbon of high resistance, made as described, and secured to metallic wires, as set forth.

2. The combination of carbon filaments with a receiver made entirely of glass and conductors passing through the glass, and from which receiver the air is exhausted, for the purposes set forth.

3. A carbon filament or strip coiled and connected to electric conductors so that only a portion of the surface of such carbon conductors shall be exposed for radiating light, as set forth.

4. The method herein described of securing the platina contact-wires to the carbon filament and carbonizing of the whole in a closed chamber, substantially as set forth.

Signed by me this 1st day of November, A. D. 1879.

THOMAS A. EDISON.

Witnesses:

S. L. GRIFFIN,

JOHN F. RANDOLPH.

3

Order of Cancellation of Certificate of Correction of Letters Patent No. 223,898.

It is found that the following certificate has been attached to Letters Patent granted to Thomas A. Edison for improvement in "Electric Lamps," No. 223,898, dated January 27, 1880:

DEPARTMENT OF THE INTERIOR,

UNITED STATES PATENT OFFICE,

WASHINGTON, D. C., *December 18, 1883.*

In compliance with the request of the party in interest Letters Patent No. 223,898, granted January 27, 1880, to Thomas A. Edison, of Menlo Park, New Jersey, for an improvement in "Electric Lamps," is hereby limited so as to expire at the same time with the patent of the following-named, having the shortest time to run, viz.: British patent, dated November 10, 1879, No. 4,576; Canadian patent, dated November 17, 1879, No. 10,654; Belgian patent, dated November 29, 1879, No. 49,884; Italian patent, dated December 6, 1879, and French patent, dated January 30, 1880, No. 133,756.

It is hereby certified that the proper entries and corrections have been made in the files and records of the Patent Office.

This amendment is made that the United States patent may conform to the provisions of section 4887 of the Revised Statutes.

[SEAL.]

BENJ. BUTTERWORTH,
Commissioner of Patents.

Approved:

M. L. JOSLYN,
Acting Secretary of the Interior.

Now, in compliance with the request of the parties in interest, said certificate is hereby canceled and proper entries and corrections have been made in the files and records of the Patent Office.

In testimony whereof I have hereunto set my hand and caused the seal of the Patent Office to be affixed, this 15th day of March, 1893.

W. E. SIMONDS,
Commissioner of Patents.

Approved:

CYRUS BUSSEY,
Assistant Secretary of the Interior.

Correction in Letters Patent No. 223,898.

DEPARTMENT OF THE INTERIOR,

UNITED STATES PATENT OFFICE,

WASHINGTON, D. C., *December 18, 1883.*

In compliance with the request of the party in interest, Letters Patent No. 223,898, granted January 27, 1880, to Thomas A. Edison, of Menlo Park, New Jersey, for an improvement in "Electric-Lamps," is hereby limited so as to expire at the same time with the patent of the following named, having the shortest time to run, viz: British Patent dated November 10, 1879, No. 4,576; Canadian Patent dated November 17, 1879, No. 10,654; Belgian Patent dated November 29, 1879, No. 49,884; Italian Patent dated December 6, 1879; and French Patent dated January 30, 1880, No. 133,756;

It is hereby certified that the proper entries and corrections have been made in the files and records of the Patent Office.

This amendment is made that the United States Patent may conform to the provisions of Section 4887 of the Revised Statutes.

BENJ. BUTTERWORTH,
Commissioner of Patents.

Approved:

M. L. JOSLYN,
Acting Secretary of the Interior.

机械电视的发明
Mechanical Television

保罗·高特列本·尼普可夫
Paul Gottlieb Nipkow

1883年，俄裔德国电气工程师尼普可夫发明了一种圆盘扫描法，它是在一个圆盘的周边，按螺旋形开若干小孔，圆盘转动时便对图像进行顺序扫描，并通过硒光电管进行电转换，实现了画像电传扫描的设想。

尼普可夫于1884年以“Elektrisches Teleskop”为名，向德国专利局提出世界上第一个关于机械电视的专利申请并获批准（DE30105）。这个专利中的尼普可夫圆盘被认为是世界上第一个电视图像光栅。但是，尼普可夫本人从来也没有做出一个模型来证明他的设计。直到1907年，放大管技术的进步才证明他的这个系统的可行性。

尼普可夫发明的圆盘扫描法被命名为“尼普可夫圆盘”，被世人称为电视的老祖宗。

DE 30105

电视机的发明
Television invention
伟大发明家贝尔德
John Logie Baird
1924年，英国人贝尔德（1988–1946年）发明了最原始的电视机。1925年，贝尔德又实现了用电传送活动图像的构想，发明了最早的电视技术。
这项发明成功后，贝尔德申请在英国开创电视广播事业，最初英国广播公司不愿意，后经议会决定才获准。1936年秋天，英国广播公司开始在伦敦播放电视节目。
贝尔德发明的第一架电视机，现被陈列在英国南肯辛顿科学博物馆中。
GB 222604
PATENT SPECIFICATION
222,604
Application Date:
PROVISIONAL SPECIFICATION
COMPLETE SPECIFICATION.
Views, Portraits, and Scenes, by Tele-
Wireless Telegraphy.
www.sipo.gov.cn

第二篇 机械电视

电视机是现代社会最常见也是最重要的电器之一，陪伴着人们的工作、学习和娱乐。我们所熟悉的“电视之父”是英国科学家约翰·洛吉·贝尔德（John Logie Baird, 1888–1946 年），他在 1925 年制造出了世界上第一台机械电视机，实现了图像、场景、动态画面的远距离传输。而贝尔德的发明是基于一种被称为“尼普可夫圆盘”的机械扫描方法。“尼普可夫圆盘”是由德国电气工程师保罗·高特列本·尼普可夫（Paul Gottlieb Nipkow, 1860–1940 年）发明的。因此，尼普可夫可谓是发明电视机的先驱。

尼普可夫发明的机械扫描法是在一个圆盘的周边，按螺旋形开若干小孔，圆盘转动时便对图像进行顺序扫描。我们可以做一个简单的试验：在一幅图像的前面放置一个这样的圆盘，圆盘前再放一块板，在板上打开一个观看的窗口。当圆盘静止时，观看的人通过窗口只能看见圆盘，看不到后面的图像；而当圆盘高速旋转时，观看的人就能通过窗口看到圆盘后面的图像了。1884 年尼普可夫在德国对他的这一发明申请了专利，并于 1885 年获得德国 DE30105 号专利权，如图 2–1、图 2–2 所示，这篇专利文献完整、清晰地披露了尼普可夫发明的机械扫描法和扫描图像的机械装置。它的原理是，当圆盘旋转起来时，每个小孔依次截取图像上各点的光信号，可以把图像中的明暗信号分开和再次结合，然后由光感器记录和传输，通过硒光电管进行电

转换，就能够复制出一幅黑白图像，从而实现图像的电传扫描。这篇专利文献中描述了传输和重现画面的过程：将图像分解为多个像素，逐个、连续传输；当多个画面快速连续出现时，在人眼中就出现了完整动态的景象——这正是电视技术的基础，直至今天，电视机仍然是按照这一基本原理工作的。

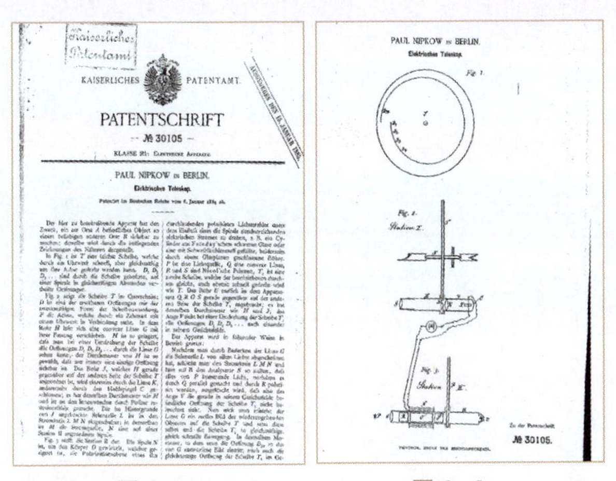

KAISERLICHES PATENTAMT.

PATENTSCHRIFT

— № 30105 —

PAUL NIPKOW in BERLIN.

Elektrisches Teleskop.

图 2-1　　　　图 2-2

尼普可夫的发明被认为是世界上第一个电视图像光栅。当时，尼普可夫将他的发明称为“电子望远镜”，而后人则以他的姓氏将这一发明命名为“尼普可夫圆盘”。但是，由于技术上的原因，尼普可夫本人未能制作出一个模型来证明他的设计，该专利也并未得到实施。直到 1907 年，放大管技术的进步才证明了尼普可夫这一发明的可行性。而发明和实现这样的电视系统的就是英国科学家贝尔德。

贝尔德在年少时就表现出对发明创造的兴趣，他曾从自己卧室到街对面的朋友家架起了一条电话线。1906 年，年仅 18 岁的贝尔德在身患疾病和生活拮据的艰苦条件下投入到对电视的潜心研制中，没有钱购买实验装置，他就亲手制作，从旧货摊、废物堆里收集各种物品，用马粪纸作出了“尼普可夫圆盘”中的旋转圆盘。在“尼普可夫圆盘”的基础上，贝尔德发明了一套能够传送图像的方法和装置。1923 年，贝尔德在英国为他的发明申请了专利，并于 1924 年

获得英国GB222604号专利权，如图2-3、图2-4所示，通过这篇专利文献我们得以了解最初的机械电视的工作原理：采用两个“尼普可夫圆盘”，一个圆盘用于发射，将景物的不同区域连续投射到光敏元件上；一个圆盘用于接收，光敏元件输出的电流通过真空管放大，被传输至接收圆盘，并点亮设置成屏幕的一系列小灯，在屏幕上通过这些小灯的明暗变化构成一幅幅图像，从而实现画面的传输和再现。

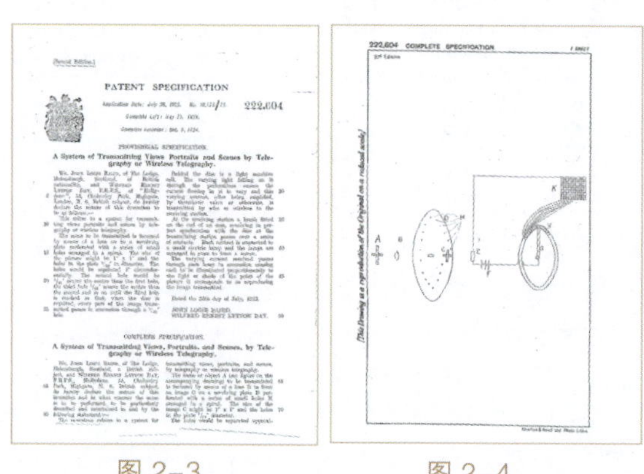

PATENT SPECIFICATION

222,604

A System of Transmitting Views Portraits and Scenes by Telegraphy or Wireless Telegraphy.

COMPLETE SPECIFICATION.

A System of Transmitting Views, Portraits, and Scenes, by Telegraphy or Wireless Telegraphy.

222,604 COMPLETE SPECIFICATION

图2-3　　图2-4

经过18年的反复实验，贝尔德在1924年第一次成功传送了一个十字剪影图像；1925年他终于成功地使人脸出现在电视机上。贝尔德的发明震惊了英国，使人们产生了收看电视广播的强烈需求，并促使英国议会最终批准了电视广播事业的开展。1929年英国广播公司与贝尔德签订专利许可合同，采用他的发明试验性地播出电视。1936年，英国广播公司利用无线电在世界上首次实现了定时电视广播。

在电视广播取得初步成功的同时，贝尔德又开始埋头研究彩色电视。另外，他的电视技术的局限性也逐步显现，画面的传送质量不高，图像不够清晰，还总是闪烁，使观众产生了很大的抱怨。1941年，贝尔德成功实现了彩色图像的传送。遗憾的是，由于劳累过度，1946年贝尔德病重离世，未能看到电视技术的进一步改进和应用。

在贝尔德根据机械扫描原理从事电视系统研究的同时，俄裔美国科学家弗拉迪米尔·科斯马·兹沃尔金（Vladimir Kosma Zowrykin）

则致力于电子扫描的研究，并开创了电视系统的另一条分支技术——电子式电视系统。

兹沃尔金先后发明了光电摄像管和光电显像管，并在此基础上发明了电子式电视。1923 年他在美国为这种电视系统申请了专利，并于 1938 年获得美国 US2141059 号专利权，如图 2–5、图 2–6 所示，从这篇专利文献中我们可以看到，这种电视系统包括一个发射站和一个接收站，都设置有包括一系列储存单元的阴极射线扫描装置——阴极射线管作为电子式电视机的核心部件。20 世纪人们普遍使用的厚重的 CRT 电视就是采用的这种技术。

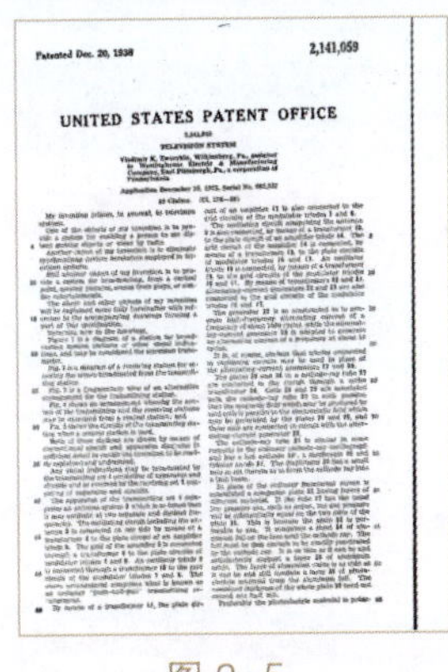

Patented Dec. 20, 1938　　2,141,059

UNITED STATES PATENT OFFICE

图 2–5

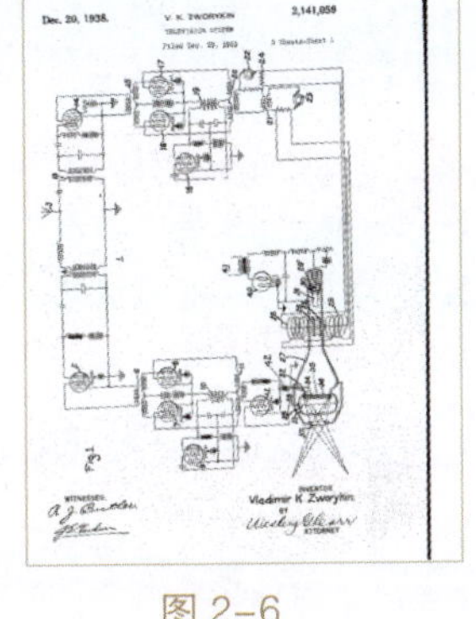

图 2–6

兹沃尔金的发明引起了美国无线电公司的关注。1928 年，美国无线电公司在纽约创建了一座实验电视台，并请兹沃尔金担任公司电子研究室主任，继续进行电视技术的开发研究。1939 年，美国无线电公司的子公司全国广播公司使用兹沃尔金的电子扫描装置第一次播出电视节目。而早在 1937 年，英国广播公司在同时试用贝尔德的机械扫描电视系统和兹沃尔金的电子扫描电视系统后，最终关闭了机械式电视系统。电子式电视系统成为主流的电视系统，并把人类带入了电视时代。

今天，电视已经成为人们生活不可缺少的一部分，电视技术更是得到了长足发展，背投、液晶、等离子、LED、3D 电视的快速更新

换代，不断给人们带来新的感官体验。回首电视技术百年的发展历程，虽然受限于当时的技术条件，尼普可夫未能真正制作出电视机，他的专利也未能获得实施，但他的发明为电视机后续的研发奠定了基石。由于尼普可夫和贝尔德的发明，机械式电视在欧洲诞生；由于兹沃尔金的发明，美国成为电子式电视的摇篮。正是有了尼普可夫、贝尔德、兹沃尔金，以及全世界众多与他们怀有同样电视梦想的发明家，通过他们的刻苦钻研和不懈努力，站在前人的肩膀上不断突破创新，电视技术才得以不断发展进步，为人类开创了一种全新的生活与沟通方式。

德国

German

1884.DE 30105

——电子望远镜

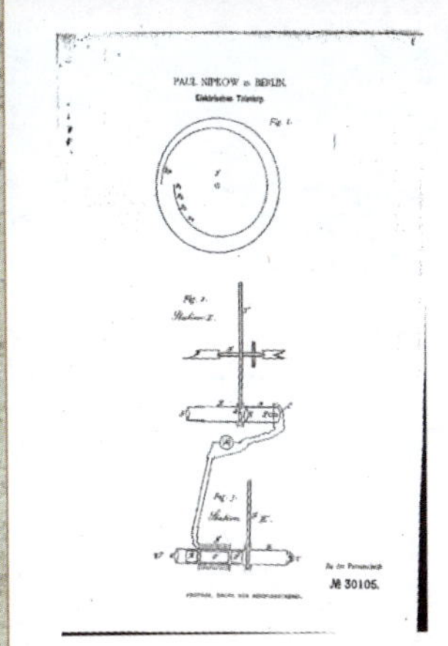

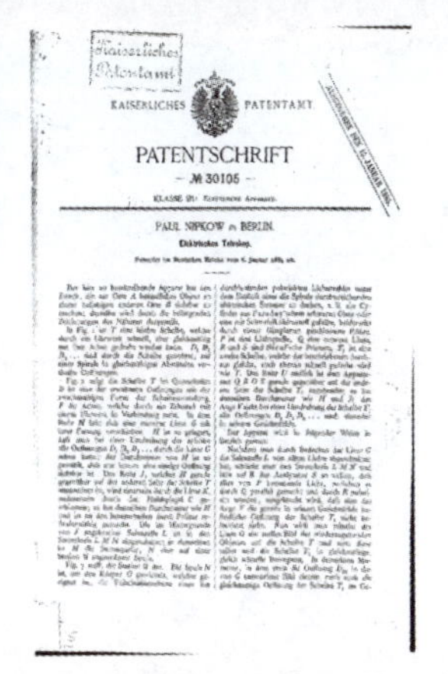
KAISERLICHES PATENTAMT

PATENTSCHRIFT

— № 30105 —

PAUL NIPKOW in BERLIN.

Elektrisches Teleskop.

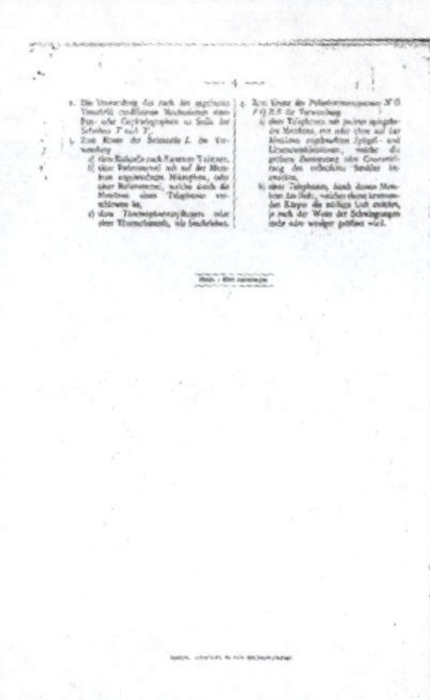

1924. GB 222604

——一种通过电报或无线电报传递肖像或景象的系统

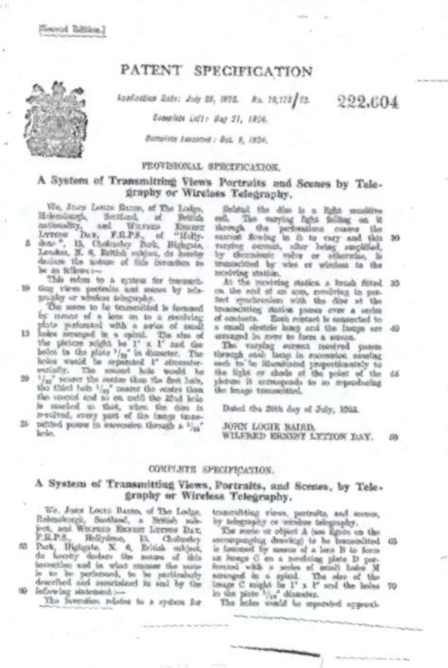

[Second Edition.]

PATENT SPECIFICATION

222,604

PROVISIONAL SPECIFICATION.

A System of Transmitting Views Portraits and Scenes by Telegraphy or Wireless Telegraphy.

Dated the 26th day of July, 1923.

JOHN LOGIE BAIRD.
WILFRED ERNEST LYTTON DAY.

COMPLETE SPECIFICATION.

A System of Transmitting Views, Portraits, and Scenes, by Telegraphy or Wireless Telegraphy.

222,604

JOHN LOGIE BAIRD.
WILFRED ERNEST LYTTON DAY.

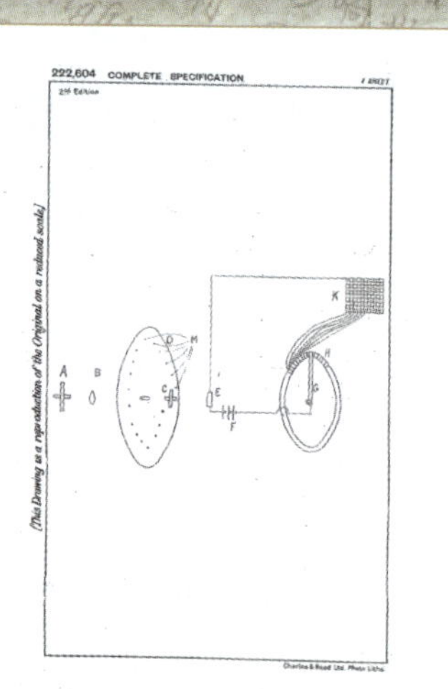

222,604 COMPLETE SPECIFICATION

[This Drawing is a reproduction of the Original on a reduced scale]

汽车的发明 Auto-mobile

卡尔·奔驰 Carl Benz

1844年，卡尔·奔驰出生于德国。中学毕业后，他上了当地一所技术学校，在学校他对机械原理特别感兴趣，尤其是偏爱研究热力发动机和蒸汽发动机。1872年退役的奔驰找朋友借钱成立了“卡尔奔驰铁器铸造和机械工厂”。于1879年12月31日晚研制成功一台单缸二行程发动机。1885年奔驰终于造出世界上第一辆汽车（单缸水冷立式发动机，排量985CC，功率0.75马力，最高车速12英里/时），并于1886年1月29日获得德国专利（DE37435），这一天被公认为世界首辆汽车诞生日。

1888年9月，奔驰的汽车在慕尼黑的机械博览会上震惊世界。1893年，卡尔·奔驰经过几年努力后，开始生产维多利亚牌四轮汽车（价格4000马克），造成产品积压，后听取商人意见推出只有2000马克的汽车，结果一年卖出125辆。

1929年，卡尔·奔驰逝世。卡尔·奔驰第一辆不用马拉的三轮车，现保存在慕尼黑的汽车博物馆。

DE 37435

第三篇 汽 车

如今的汽车，同我们所熟知并依赖的其他现代科技工具一样，起源于一个历史时间点并经历了漫长的发展历程。16 世纪初第一次工业革命后，基于动力核心技术的重大变革，以蒸汽为动力驱动的交通工具开始出现，但都因未能突破笨重及低效的弊端而停止了发展。

1879 年 12 月，德国工程师卡尔·奔驰（Carl Benz，1844–1929 年），首次试验成功了一台单缸煤气发动机。此后，奔驰又在 1886 年研制成功了单缸汽油发动机，发明了第一辆不用马拉的三轮车，并于 1886 年 1 月 29 日获得德国 DE37435 号专利权。最初的汽车，其实只是一辆不靠马拉的马车，技术上也需要继续磨合，因此并未得到广泛认可。直到 1888 年 9 月，奔驰发明的汽车在慕尼黑博览会上取得了非常大的轰动，并接到了大量订单。自此，奔驰公司开始蓬勃发展。1893 年，在一度滞销后，降低成本的维多利亚牌四轮汽车以 125 辆的年度销售量再次令奔驰汽车备受瞩目。1901 年，我国清朝皇室拥有的第一辆汽车就是奔驰汽车。1929 年，奔驰离世。作为现代汽车工业的先驱者之一，卡尔·奔驰被人们誉为“汽车之父”。

奔驰的第一项“汽车”德国 DE37435 号专利保护了一个由燃气发动机驱动的交通工具的动力系统，以及制动系统。这款汽车是在三轮车的底部框架后下方增加了动力装置，从而实现了驱动力上的突破，彻底解放了人力、畜力，如图 3–1、图 3–2、图 3–3 所示。

KAISERLICHES PATENTAMT.

PATENTSCHRIFT

№ 37435

BENZ & CO. in MANNHEIM.

Fahrzeug mit Gasmotorenbetrieb.

图 3-1

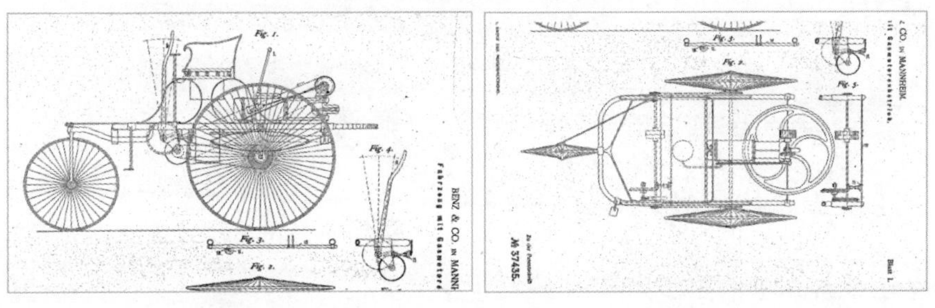

图 3-2　　图 3-3

一百多年过去了，汽车已经成为普及性的交通工具，虽然无论从性能还是外形上，都经历了跨越式的变革，但奔驰的汽车发明依然在现代汽车工业中发挥着启迪作用。通过专利文献检索我们发现，2012 年一项名为“排座型全地形车辆”的发明专利依然是参考了奔驰的发明。该发明由美国北极星工业有限公司申请，为在尽可能多的国家受到保护，该公司提交了国际专利申请（WO2012109546）。作为技术保护目标地区之一，中国也在该国际申请指定的国家范围之内，如图 3-4、图 3-5、图 3-6 所示。

美国北极星工业有限公司是当今一家全地形车辆制造商，主要生产雪地行走专用车、美式巡航摩托车等，也是美国军方重要的轻型作战车辆供应商。从专利文献中我们可以看到卡尔·奔驰发明的汽车排

座、后驱动等技术点仍在这项专利中得到应用。这也意味着，100 多年前人类的智慧结晶，在现今的人类社会中依然发挥着效能。

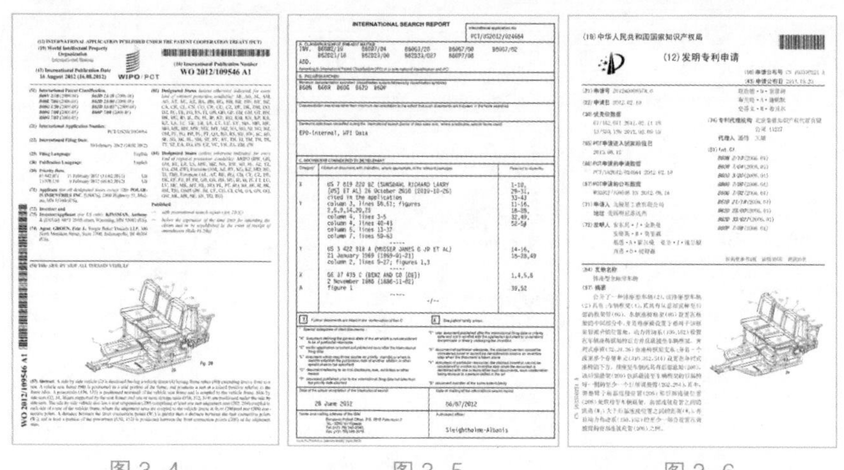
WO 2012/109546 A1

INTERNATIONAL SEARCH REPORT

(12) 发明专利申请

图 3-4　　图 3-5　　图 3-6

卡尔·奔驰创办的奔驰汽车公司（现名为戴姆勒 - 奔驰汽车公司），如今是全世界最著名的汽车公司之一。曾经有人这样评论卡尔·奔驰："世界公认的汽车发明者，以其非凡的才智和坚忍不拔的钻研精神，创造出了汽车这一令世界惊叹的交通工具，从而提高了现代人的生活质量，拓展了现代人的生活空间。"

1886.DE 37435

——车用天然气发动机

KAISERLICHES PATENTAMT.

PATENTSCHRIFT

№ 37435

KLASSE 46: LUFT- UND GASKRAFTMASCHINEN.

BENZ & CO. IN MANNHEIM.

Fahrzeug mit Gasmotorenbetrieb.

Patentirt im Deutschen Reiche vom 29. Januar 1886 ab.

德国
German

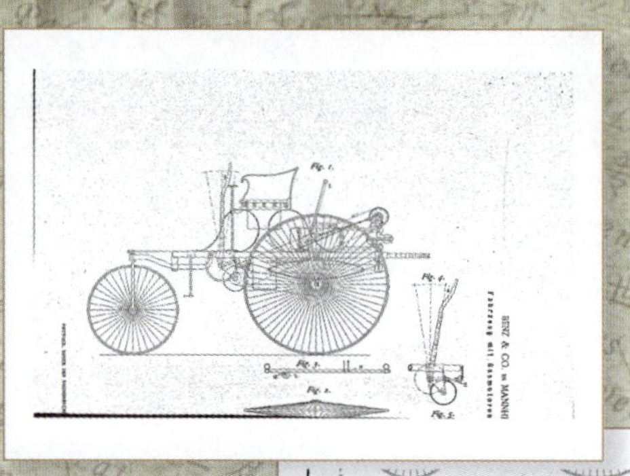

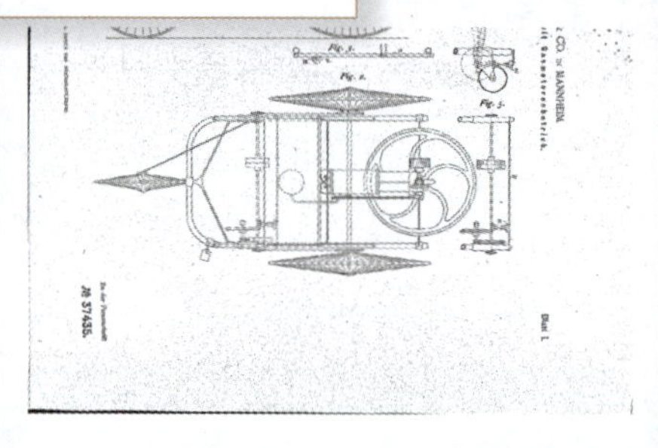
№ 37435.

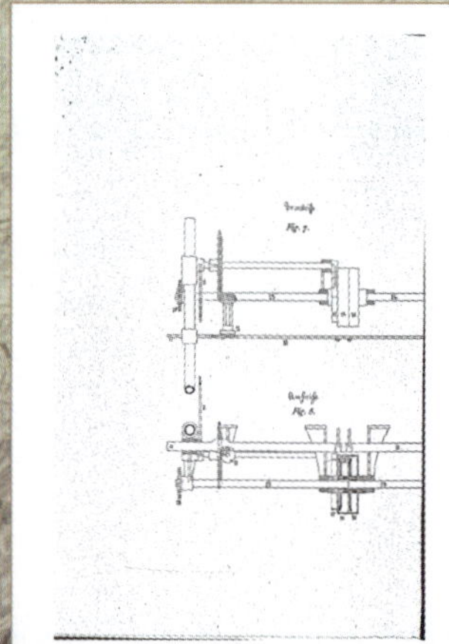

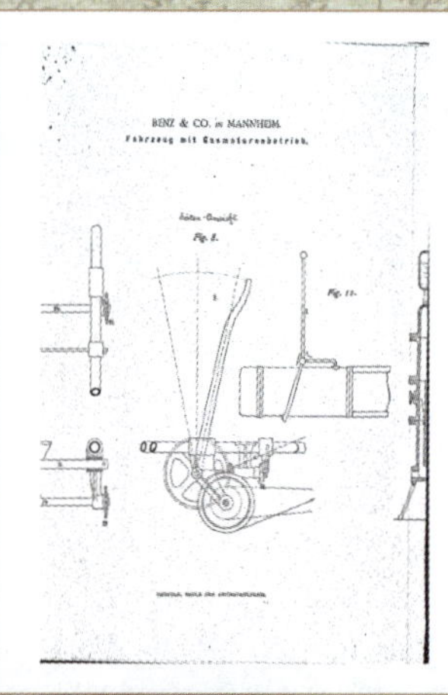
BENZ & CO. in MANNHEIM.
Fahrzeug mit Gasmotorenbetrieb.

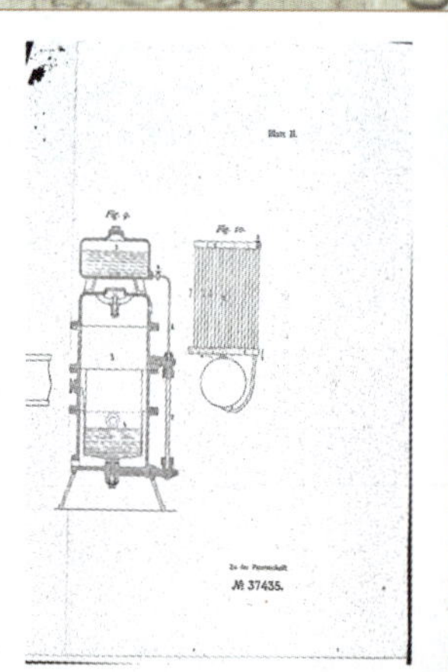
Blatt II.
№ 37435.

洗碗机的发明
Dish Washing Machine

约瑟芬·考克兰
Josephine Cochrane

约瑟芬·考克兰，美国人，世界早期女性发明家。她于1886年发明了第一台实用机械洗碗机。

约瑟芬·考克兰（1839–1913年），出生于美国伊利诺伊州的谢尔比维尔。作为贵妇约瑟芬·考克兰视名贵瓷器如珍宝，为了减少家宴中瓷器的损伤，她萌发了创造安全的机械洗碗机的想法。约瑟芬·考克兰在自家木棚里建造了简易车间，试验成功了洗碗机。由于约瑟芬发明的洗碗机具有很高的实用性，伊利诺伊州的酒店、宾馆纷纷要求订货，约瑟芬·考克兰决定将洗碗机投入市场。

1886年，约瑟芬为自己的发明申请了专利，同年12月28日获得一项美国专利 。

US 355139

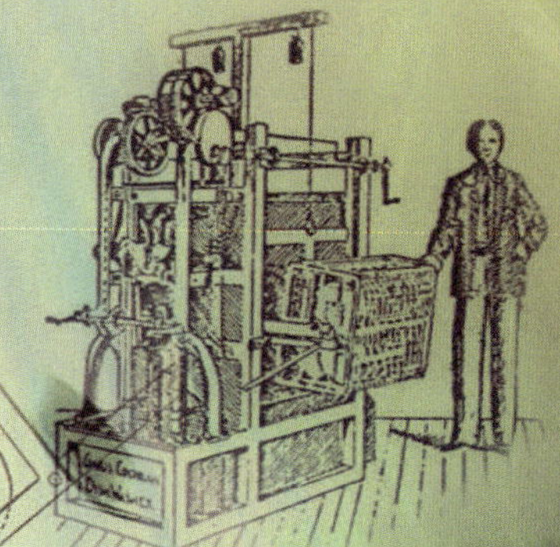

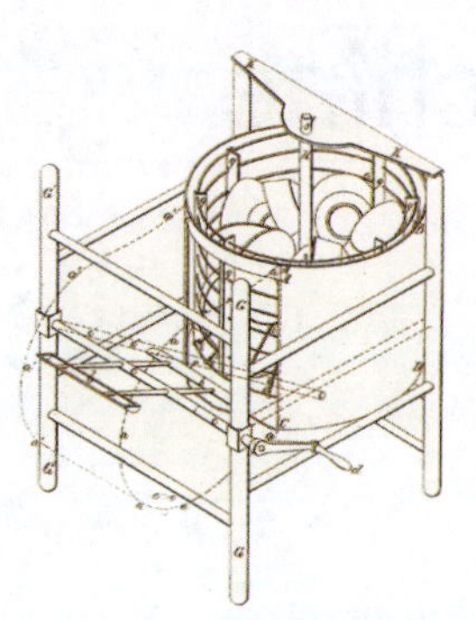

第四篇 洗碗机

洗碗是人类日常生活中不可或缺的一项劳动。随着科学技术和工艺水平的进步，人们开始尝试利用机械工具辅助洗碗。洗碗机的出现极大地减轻了人们厨房劳动的负担，无论在餐厅还是家庭，它都使洗碗变得更加便捷和高效，给人们带来了整洁、卫生和轻松。

1850 年美国人乔尔·霍顿（Joel Houghton）发明了一台“桌面家具清洁机”，用于清洗桌面上摆放的小家具，特别是餐具，因此被人们理解为是一台“洗碗机”。这项发明获得了美国 US7365 号专利权，如图 4-1、图 4-2 所示，这台机器用木头制成，通过手摇一个轮盘将水冲到餐具的表面进行清洗。然而，在实践中，这台机器几乎不能使用——费力却并不能达到将餐具冲洗干净的目的。虽然该项发明

UNITED STATES PATENT OFFICE.

JOEL HOUGHTON, OF OGDEN, NEW YORK.

IMPROVEMENT IN MACHINES FOR WASHING TABLE FURNITURE.

图 4-1

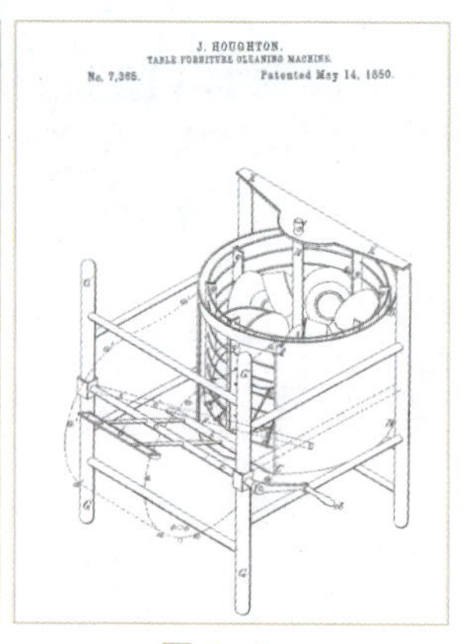

图 4-2

的初衷是希望利用机械设备取代手工洗碗，但实际上未能取得成功。

在这项发明后不久，世界公认的第一台真正的洗碗机问世了，它的发明者是一位美国女性，她就是被称为现代洗碗机之母的约瑟芬·M. 加里斯·考克兰（Josephine Cochrane，1839–1913 年）。考克兰生长于美国伊利诺伊州一个富裕的家庭，她的祖父是蒸汽船的发明者约翰·菲奇（John Fitch），她的父亲是一位土木工程师。考克兰家中拥有不少名贵的瓷器餐具，由于经常设宴招待当地社会名流，瓷器餐具需要经常使用和清洗。仆人在刷洗时总是难以避免地会摔坏或碰坏餐具，这让视瓷器如珍宝的考克兰倍感心疼。考克兰从小就受到父辈们创新精神和技术才干的影响与启迪，她萌发了创造一种安全的洗碗机的想法。考克兰在自家的木棚里搭建了简易车间，经过反复试验，终于成功发明了洗碗机。1885 年考克兰将她的发明在美国申请了专利，1886 年获得名为“洗碗机”的美国 US355139 号专利权，如图 4–3、图 4–4、图 4–5 所示。

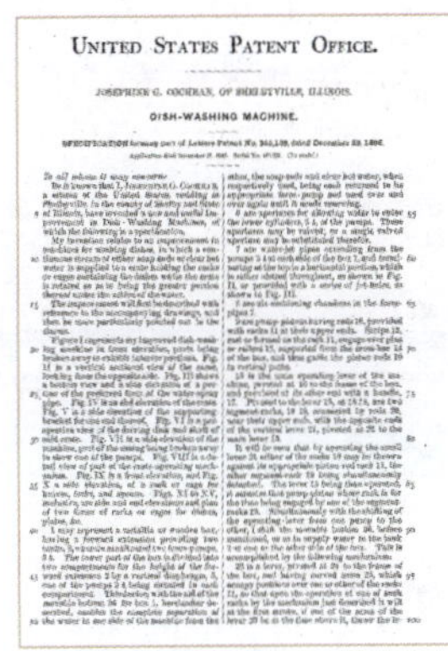

UNITED STATES PATENT OFFICE.

DISH-WASHING MACHINE.

图 4–3

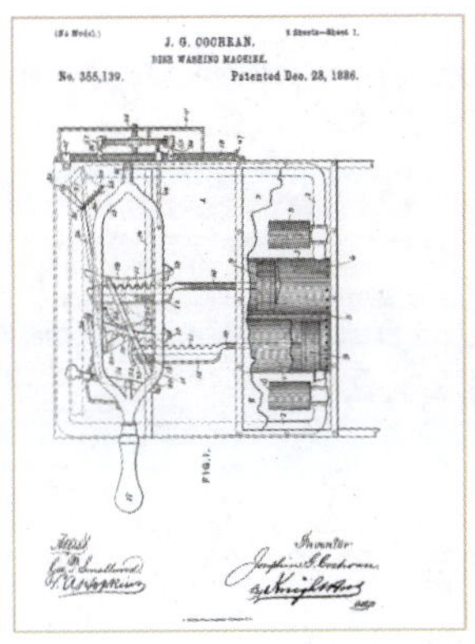

图 4–4

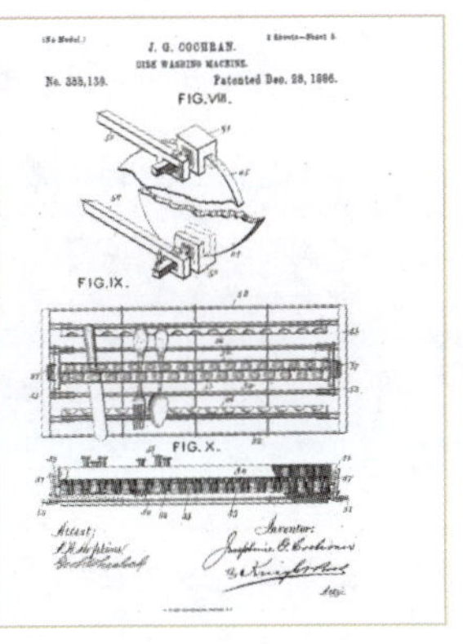

图 4–5

通过这篇专利文献，我们得以了解考克兰的发明：它是用金属丝弯制成盛装碗碟用的框架（或称碗笼），将该框架装在一个转轮上，然后整体放置在热水箱里。摇动手柄，以手柄带动碗架转动，使水箱里的水泵喷出热水，热水随即喷洒在转动的碗碟上。在洗碗机内加装了混有热肥皂水的喷嘴，来喷淋支架上的餐具。从考克兰发明的背景

和技术方案可以看出，她首要解决的是保证碗碟不被损坏的问题，其次才是节省人力。

1893 年，在规模盛大的芝加哥世界博览会上，考克兰的洗碗机首次登场展出，立刻得到了广泛的赞赏和欢迎。此后的几年，考克兰经历了丈夫离世、家道中落、背负债务一系列变故，在生活压力下她产生了把洗碗机投入市场的想法。1897 年，考克兰创立了加里斯 · 考克兰制造公司，专门生产洗碗机。在考克兰去世后，该公司被豪霸公司收购。豪霸公司后来生产的 KitchenAid 牌洗碗机（目前隶属于惠而浦集团）主要供应宾馆和餐馆，直到 20 世纪 60 年代中期以后，洗碗机才逐渐走入寻常百姓家。

1949 年，美国通用电气公司在美国提交了洗碗机专利申请并于 1953 年获得了美国 US2654377 号专利权，如图 4-6、图 4-7 所示，从这篇专利文献中可以清晰地看到，这是一台圆桶形的电动洗碗机，并通过软管连接水龙头来输水。这台电动台式洗碗机无论是外形还是使用便捷程度都较手动洗碗机有了显著的进步，为洗碗机的普及奠定了基础。1954 年，通用公司生产出第一台电动台式洗碗机。这台洗碗机不仅提高了洗涤性能，还改善了整机体积外形。

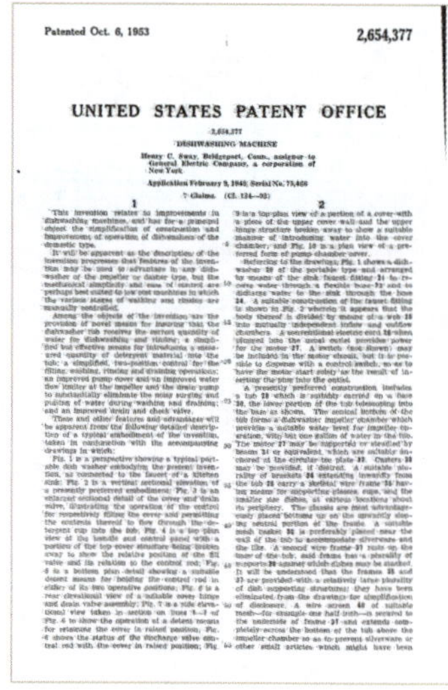

Patented Oct. 6, 1953 2,654,377

UNITED STATES PATENT OFFICE

DISHWASHING MACHINE

图 4-6

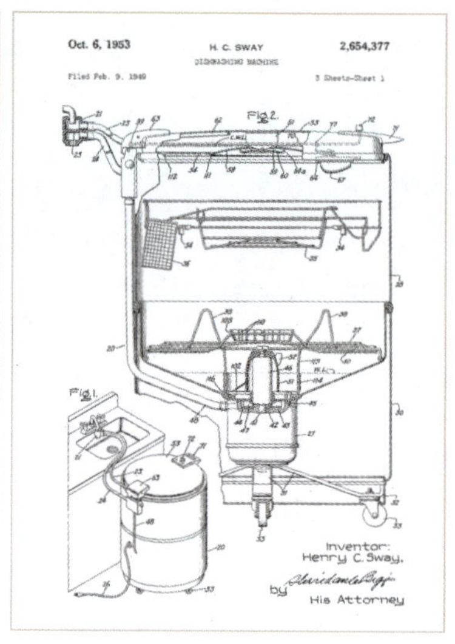

图 4-7

1951 年通用公司提交了另一项可嵌入工作台下的洗碗机专利申请，并于 1953 年获得美国 US2661750 号专利权，如图 4-8、图 4-9 所示。这正是我们今天厨房常见的嵌入式洗碗机。

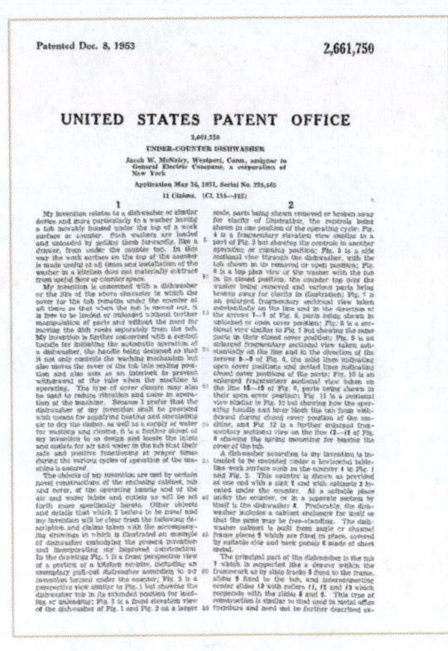

Patented Dec. 8, 1953 2,661,750

UNITED STATES PATENT OFFICE

UNDER-COUNTER DISHWASHER

Jacob W. McNairy, Westport, Conn., assignor to General Electric Company, a corporation of New York

图 4-8

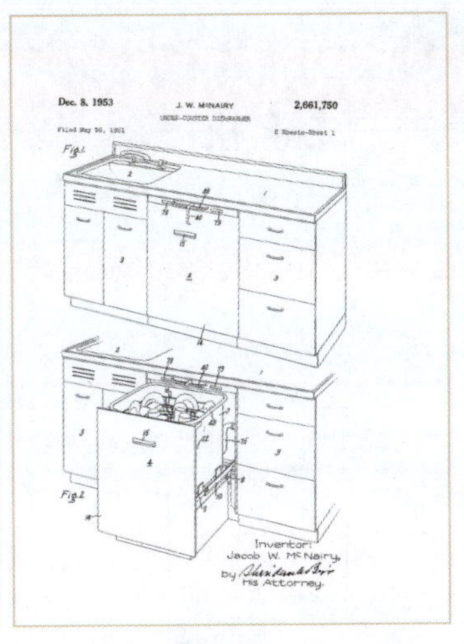

图 4-9

在此后的 40 多年时间中，随着微电脑控制技术的引入，洗碗机变得更加智能和人性化，真正成为人们在厨房里的亲密帮手。今天，世界各地的家庭主妇在享受洗碗机带来的整洁、方便和轻松时，都应当感谢洗碗机的发明者考克兰。考克兰发明洗碗机的故事也告诉我们，自己动手、运用智慧解决生活中的问题，就有可能改变我们的生活。

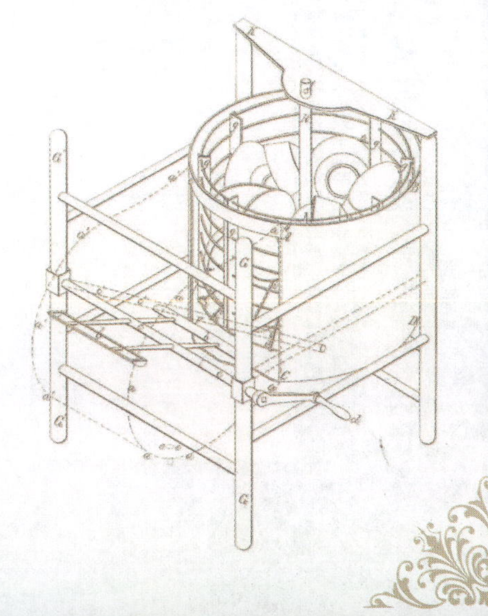

美国 America

1886.US 355139

洗碗机

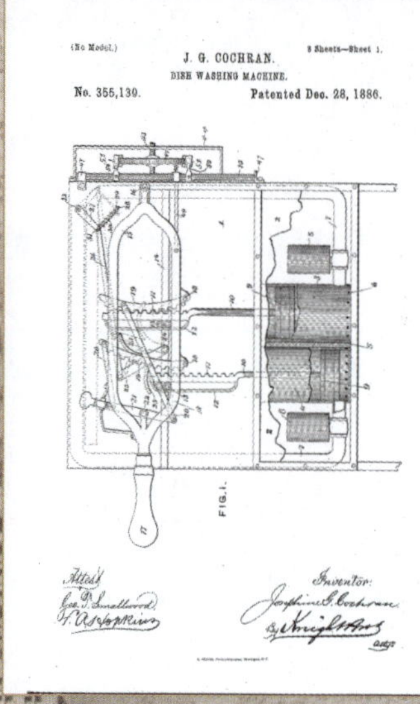

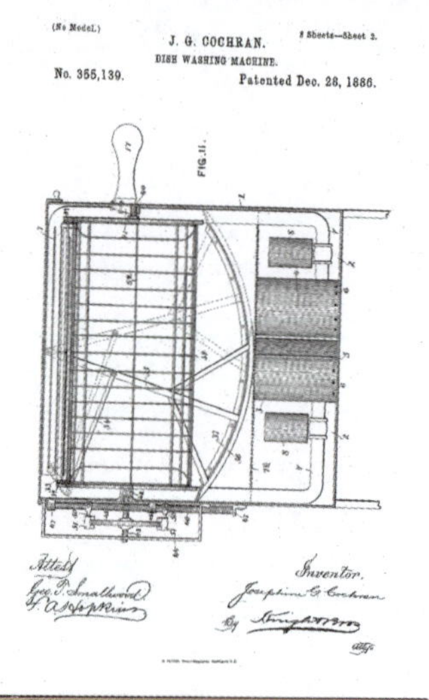

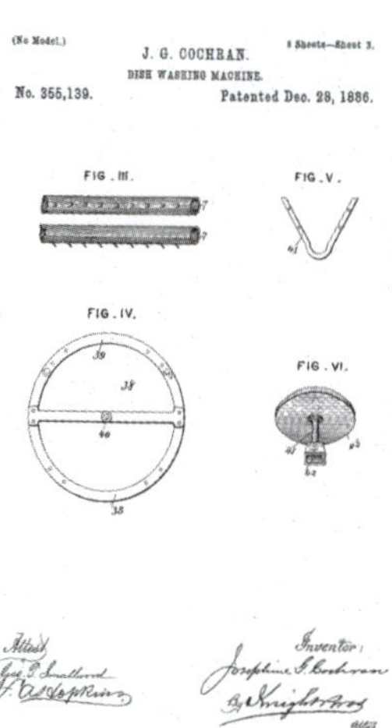

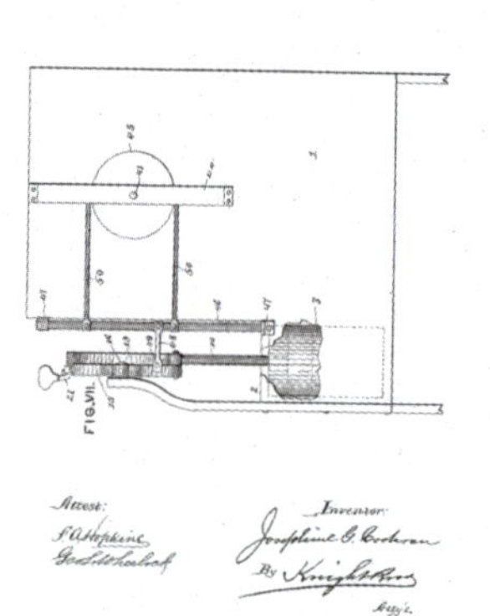

美国 America

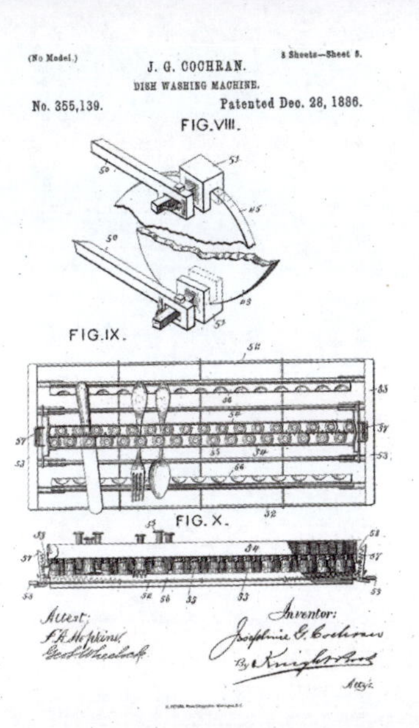

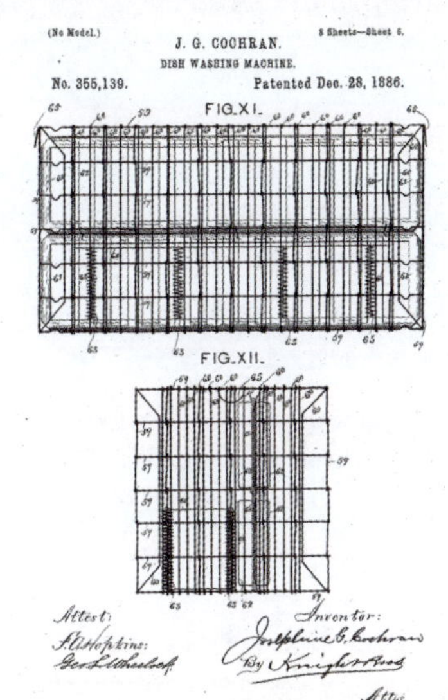

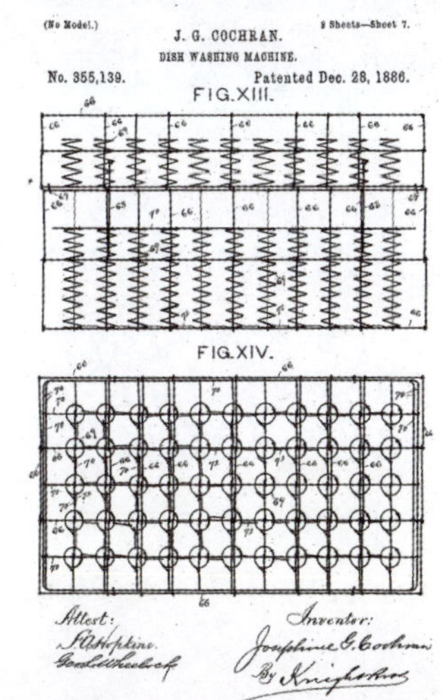

(No Model.)
J. G. COCHRAN.
5 Sheets—Sheet 8.
DISH WASHING MACHINE.
No. 355,139.
Patented Dec. 28, 1886.
FIG.XV.

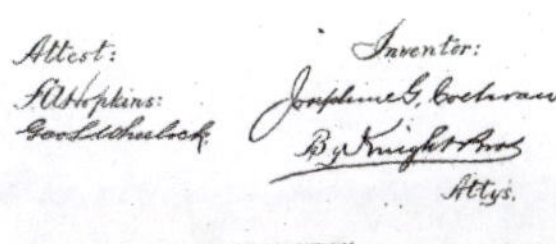

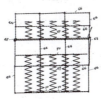

美国 America

UNITED STATES PATENT OFFICE.

JOSEPHINE G. COCHRAN, OF SHELBYVILLE, ILLINOIS.

DISH-WASHING MACHINE.

SPECIFICATION forming part of Letters Patent No. 355,139, dated December 28, 1886.

Application filed December 31, 1885. Serial No. 197,178. (No model.)

To all whom it may concern:

Be it known that I, JOSEPHINE G. COCHRAN, a citizen of the United States, residing at Shelbyville, in the county of Shelby and State of Illinois, have invented a new and useful Improvement in Dish-Washing Machines, of which the following is a specification.

My invention relates to an improvement in machines for washing dishes, in which a continuous stream of either soap-suds or clear hot water is supplied to a crate holding the racks or cages containing the dishes while the crate is rotated so as to bring the greater portion thereof under the action of the water.

The improvement will first be described with reference to the accompanying drawings, and then be more particularly pointed out in the claims.

Figure I represents my improved dish-washing machine in front elevation, parts being broken away to exhibit interior portions. Fig. II is a vertical sectional view of the same, looking from the opposite side. Fig. III shows a bottom view and a side elevation of a portion of the preferred form of the water-spray pipe. Fig. IV is an end elevation of the crate. Fig. V is a side elevation of the supporting-bracket for one end thereof. Fig. VI is a perspective view of the driving disk and shaft of said crate. Fig. VII is a side elevation of the machine, part of the casing being broken away to show one of the pumps. Fig. VIII is a detail view of part of the crate-operating mechanism. Fig. IX is a front elevation, and Fig. X a side elevation, of a rack or cage for knives, forks, and spoons. Figs. XI to XV, inclusive, are side and end elevations and plan of two forms of racks or cages for dishes, plates, &c.

1 may represent a metallic or wooden box, having a forward extension providing two tanks, 2, wherein are situated two force-pumps, 3 4. The lower part of the box is divided into two compartments for the height of the forward extension 2 by a vertical diaphragm, 5, one of the pumps 3 4 being situated in each compartment. This device, with the aid of the movable bottom 36 for box 1, hereinafter described, enables the complete separation of the water in one side of the machine from the other, the soap-suds and clear hot water, when respectively used, being each returned to its appropriate force-pump and used over and over again until it needs renewing.

6 are apertures for allowing water to enter the lower cylinders, 3 4, of the pumps. These apertures may be valved; or a single valved aperture may be substituted therefor.

7 are water-jet pipes extending from the pumps 3 4 at each side of the box 1, and terminating at the top in a horizontal portion, which is either slotted throughout, as shown in Fig. II, or provided with a series of jet-holes, as shown in Fig. III.

8 are air-cushioning chambers in the force-pipes 7.

9 are pump-pistons having rods 10, provided with racks 11 at their upper ends. Strips 12, cast or formed on the rack 11, engage over pins or rollers 13, supported from the cross-bar 14 of the box, and thus guide the piston-rods 10 in vertical paths.

15 is the main operating-lever of the machine, pivoted at 16 to the frame of the box, and provided at its other end with a handle, 17. Pivoted to the lever 15, at 18 18, are two segment-racks, 19 19, connected by rods 20, near their upper ends, with the opposite ends of the vertical lever 21, pivoted at 22 to the main lever 15.

It will be seen that by operating the small lever 21 either of the racks 19 may be thrown against its appropriate piston-rod rack 11, the other segment-rack 19 being simultaneously detached. The lever 15 being then operated, it actuates that pump-piston whose rack is for the time being engaged by one of the segment-racks 19. Simultaneously with the shifting of the operating-lever from one pump to the other, I shift the movable bottom 36, before mentioned, so as to supply water to the tank 2 at one or the other side of the box. This is accomplished by the following mechanism:

23 is a lever, pivoted at 24 to the frame of the box, and having curved arms 25, which occupy positions over one or other of the racks 11, so that upon the operation of one of such racks by the mechanism just described it will at the first stroke, if one of the arms of the lever 23 be at the time above it, throw the le-

2

355,139

ver 23 over its fulcruming-point 24, and thus allow said rack to reciprocate without again coming in contact with the arm 25.

26 is a rod connecting the lever 23 with a crank, 27.

28 is a pin or bolt hinged to the connection between the rod 26 and crank 27 at its upper end, and sliding in a bracket, 29, on the box at its lower end. A spring, 30, surrounding the pin 28 bears on the bracket 29 and crank 27, so as to insure the throwing of such crank, and consequently of the lever 23, to its extreme position as soon as the knuckle-joint at 31 passes to one or the other side of a straight line drawn between the pivot-pin 32 of crank 27 and the center of bracket 29. The pivot-rod 32 is rigidly connected interiorly of the box to a second crank, 33, and thus, by connecting rod 34 with arms 35 of the movable bottom 36, a ledge, 37, is fixed to each side of the box, immediately over the path of the movable bottom 36, so as to direct water falling down the sides of the box onto the bottom.

From the above-described construction it is apparent that as soon as one or other of the pumps is operated the movable bottom is automatically shifted from the appropriate side, so that water is returned to the tank of the pump being used and warded off from the other tank.

An oil or other stove or heater may be placed under one or other of the tanks 2 2, which it is preferred to employ for containing the hot water.

38, Figs. II and IV, is a cylindrical wire crate, wherein the cages containing the dishes to be cleaned are placed.

39 is a door at one side of the crate, having suitable means of fastening. (Not here shown.) A circular trunnion, 40, at one end of the crate 38 is supported in a bracket, 41, Fig. V, which is fixed to the side of the box and open at top, so as to allow the crate to be readily removed. At its other end the crate has a square projection or trunnion for engaging in a similarly-shaped socket, 42, Fig. VI, on the inner end of a shaft, 43, Fig. II, journaled in the wall of the box and in a bracket, 44.

45 is a friction-disk fixed on the shaft 43 so as to revolve in a vertical plane.

46 is a vertical rock-shaft journaled in brackets 47, and having a horizontal crank-arm, 48, connected by rod 49 with the main operating-lever 15, and having additional horizontal crank-arms, 50, carrying friction-clutches 51. The said clutches engage with opposite sides of the disk 45 in one direction of their motion and not the other, so as to alternately grasp and rotate the same.

During the continuous operation of the lever 15, and while one of the pumps 3 4 is supplying a steady stream of water through the pipes 7 to dishes in the crate 38, the shaft 46 is by its connection with the lever 15 rocked, and the disk 45 and crate 38 slowly rotated, so that all the dishes in the crate are exposed to the stream of water and cleaned. After they are thoroughly cleaned by soap-suds from one pump, the racks 11 19 are shifted so as to bring the other clear-hot-water pump into operation, thus automatically uncovering the proper tank in the bottom of the box. By this means a stream of hot rinsing-water is poured onto the dishes until they are perfectly clean. When they are dry from their own heat they are ready to be put away or immediately again used upon the table, no draining or drying being necessary after their removal from the machine.

To strain off the larger impurities a perforated diaphragm may be placed under the movable bottom of the tank, as shown in Fig. II at 72.

A rack or cage for knives and similar articles is shown in Figs. IX and X. It consists of a frame, 52, preferably of wire, to the end plates, 53, of which are secured the ends of the bars 54, which support a number of spring-clamps, 55. Articles to be washed are laid upon the bars 56 of the frame, and the clamp-bars 54 are then placed over them and are held down by the racks 57, engaging with projections 58 on the ends of the bars 54.

In Figs. XI and XII are shown side and end elevations of a rack for plates, saucers, &c. The outer web of the cage is made of wires 59, so as to be practically rectangular. From each corner wire projects into the case a series of wires, 60, which for the major part of their length are parallel with the outer wires of the cage, and which serve as supports for the plates placed in the cage in vertical position. Frames 61, having vertical wires 62, which may project, as shown, and may be similar to those 60, may be placed between any two of the wires 60 at the ends of the cage, and thus divide the cage into two parts, and at the same time afford an interior bearing for the plates and saucers that are placed within the cage. For articles that require a spring support below, an interior cushion is provided in one of the compartments of the cage, consisting of springs 63, united by longitudinal and cross bars 64. Handles 65 are provided at each end of the cage for convenience in carrying.

Figs. XIII, XIV, and XV are respectively side elevation, plan, and end elevation of a rack for glass and other hollow articles. The rack or cage is made of wires 66, and is divided into two parts which are hinged together, as shown at 67; but they may be fixed so as to form a rectangular cage by means of catches 68. Within each compartment of this cage a number of spiral springs, 69, are arranged upon a movable frame, 70, and so adapted as to surround the necks and bases of hollow articles, and by pressing them gently against the top and bottom of the cage hold them in position while the cage is being turned within the crate of the dish-washing machine. To increase their power of maintaining the dishes in upright position, the lower ends of

355,139

3

the springs 69 are preferably connected by links 71.

Having thus described my invention, the following is what I claim as new therein and desire to secure by Letters Patent:

1. In combination with the box of a dish-washing machine, receptacles or tanks connected therewith and a movable bottom adapted to cover one receptacle and deflect water falling thereon into the other, substantially as set forth.

2. In combination with the box of a dish-washing machine, an operating-lever, pumps, racks on the piston-rods of said pumps, pivoted racks supported on the main operating-lever, and means for throwing one or other of said pivoted racks into engagement with the rack on one of the piston-rods, substantially as set forth.

3. In combination with the box of a dish-washing machine, partitioned as shown, two pumps, means of connecting one or other at will with the operating handle or lever, a movable bottom for the box, and connections therefrom to the pump pistons, whereby it is automatically shifted as each pump is alternately operated, substantially as set forth.

4. In combination with the box 1, the pump piston-rods and a lever for operating the same, a lever, 23, pivoted to the box, having arms 25, projecting over the piston-rods, and a movable bottom wherewith said lever 23 is connected, substantially as set forth.

5. In combination with the lever 23 and means for shifting the same, a crank, 27, a rod, 26, connecting said crank and lever, a movable bottom operated by said crank, and a spring-bolt, 28 29 30, for throwing said crank and lever to its extreme positions to one side or the other, substantially as set forth.

6. The combination, with a box and rotary crate, of two pumps having independent discharge-pipes, a hinged lever adapted to engage either pump-rod, and a connection, substantially as described, by which the lever rotates the crate while connected with either pump.

7. In combination with a horizontal rotary dish-crate, a horizontal shaft connected thereto, a friction-disk thereon, a vertical rock-shaft carrying friction-clutches engaging with said disk, and an operating-lever connected with said vertical rock-shaft, substantially as set forth.

8. In combination with a rotary dish-crate having a square trunnion at one end for engaging with the operating-shaft and a circular trunnion at the other, a bracket fixed to the box having bearing for the circular trunnion, open at top to permit the removal of the crate, substantially as set forth.

9. A knife-and-fork rack or cage for dish-washing machines, consisting of a frame, 52, provided with end plates, 53, having racks 57, the bars 54, having projections 58, the bars 56, on which the articles are placed, and spring-clamps 55, substantially as described.

10. A plate-and-saucer rack or cage for dish-washing machines, consisting of an approximately rectangular wire cage having an outer web of wires, 59, and the inwardly-projecting vertical wires 60, extending from the corners around the sides and ends, substantially as described.

11. A plate-and-saucer rack or cage for dish-washing machines, consisting of an approximately rectangular wire cage having an outer web of wires, 59, the inwardly-projecting vertical wires 60, extending from the corners around the sides and ends, and the interior frames, 61, having vertical wires 62, similar to the inwardly-projecting vertical wires, substantially as described.

12. A rack for hollow articles, consisting of two wire cages having wires 66, hinged together, means for securing them together at the front, and movable frame 70, spiral springs 69, and connecting-links 71, substantially as described.

JOSEPHINE G. COCHRAN.

Witnesses:

J. WM. LLOYD,

R. L. GARIS.

交流发电机的发明

Alternating Current Electro-Magnetic Motor

尼古拉·特斯拉

Nikola Tesla

尼古拉·特斯拉（1856–1943 年），美籍南斯拉夫人，发明家、物理学家、机械工程师和电机工程师。他一生的发明无数。

1882 年，继爱迪生发明直流电（DC）后不久，特斯拉就发明了交流电（AC），并于 1888 年 5 月获得了美国专利（US381968），制造出世界上第一台交流发电机。1895 年，他替美国尼亚加拉发电站制造发电机组，该发电站至今仍是世界著名水电站之一。1897 年，他使马可尼的无线电通信理论成为现实。1898 年，他制造出世界上第一艘无线电遥控船，无线电遥空技术取得专利（US613809）。1899 年，他发明了 X 光摄影技术。

1956 年是特斯拉诞辰一百周年，国际电气技术协会决定，把国际单位制中磁感应强度的单位命名为特斯拉，简称“特”

US 381968

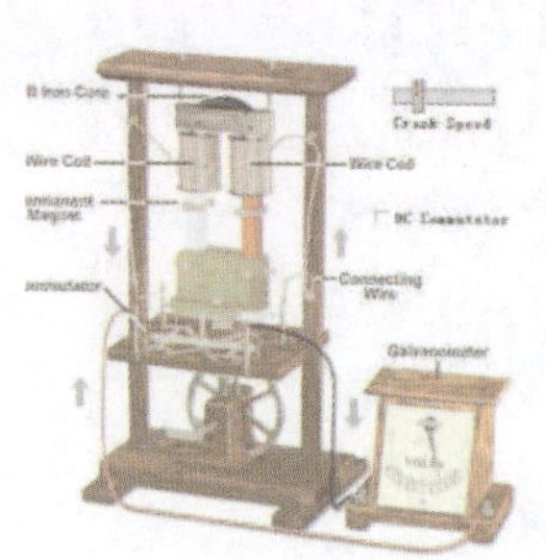

第五篇 交流发电机

电能，是人类现代社会不可缺少的能源。1831 年，迈克尔·法拉第（Michael Faraday）将一个封闭电路中的导线通过电磁场，导线转动有电流过电线，人类从此了解到电和磁场之间有某种紧密的关联。随后，法拉第建造了第一座发电机原型，由此产生了电力。1880 年，爱迪生研制出直流发电机，1882 年爱迪生电气照明公司在纽约建立了第一座发电站，电能开始在工农业生产和日常生活等方面得以应用。

为了保证远距离输送电的效率，人们开始思考利用电压容易改变的“交流电”。被誉为交流电之父、无线电之父的美国机械工程师、电机工程师尼古拉·特斯拉（Nikola Tesla，1856–1943 年）在 1887 年组装完成了世界上最早的无电刷交流电感应马达。同年，特斯拉为他发明的电磁发电机在美国申请了专利，并于 1888 年获得名

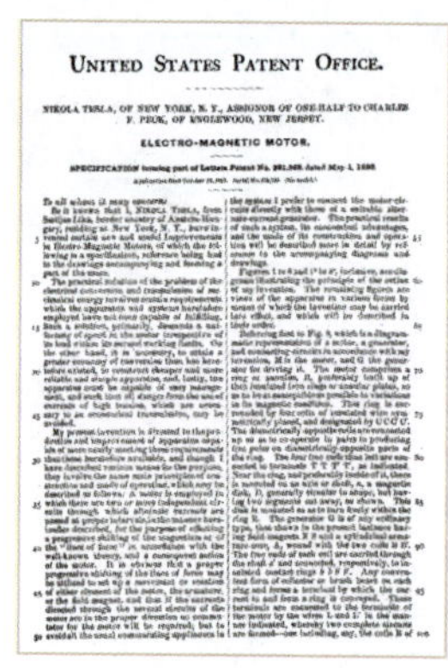

UNITED STATES PATENT OFFICE.

NIKOLA TESLA, OF NEW YORK, N. Y., ASSIGNOR OF ONE-HALF TO CHARLES F. PECK, OF ENGLEWOOD, NEW JERSEY.

ELECTRO-MAGNETIC MOTOR.

图 5-1

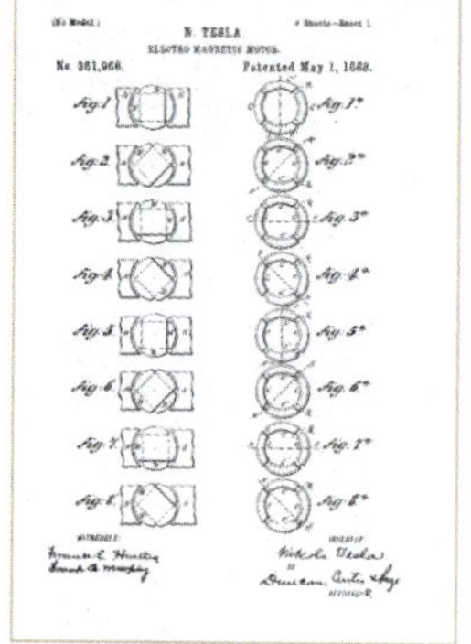

N. TESLA
ELECTRO MAGNETIC MOTOR.
No. 381,968. Patented May 1, 1888.

图 5-2

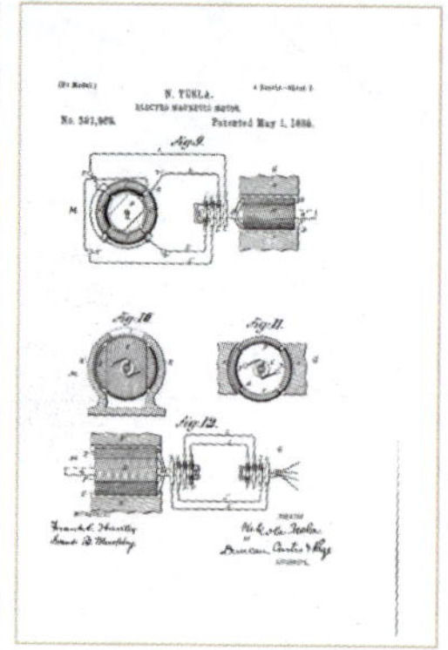

N. TESLA.
ELECTRO MAGNETIC MOTOR.
No. 381,968. Patented May 1, 1888.

图 5-3

为“电磁电机”的美国 US381968 号专利权，如图 5-1、图 5-2、图 5-3 所示。

这件专利所保护的是一种由电枢（发电机的定子）、线圈磁极（发电机的转子）和发电机形成的交流发电机。定子由电枢铁芯、均匀排放的三相绕组及机座和端盖等组成。转子的线圈磁极通入直流电流，产生接近于正弦分布磁场（称为转子磁场），其有效励磁线圈与静止的电枢绕组相交链。转子旋转时，转子磁场随同一起旋转、每转一周，磁力线顺序切割定子的每相绕组，在三相定子绕组内感应出三相交流电势，从而提供更安全、可靠和简单的发电系统。

此后多年，特斯拉专注于交流电电力系统和电力传输技术的研究，先后共申请了 60 余项相关专利，如图 5-4 所示。其中，不仅包括交流电产生装置，还包括电力传输系统、稳压装置、交流电分流装置、交流电电流接收装置和交流电流量测试装置等。这一系列专利奠定了交流电电力系统的基础，成为人类社会的宝贵财富。值得一提的是，特斯拉于 1890 年就提出了关于“高频率”（15000 赫兹）交流发电机的专利申请，于 1891 年获得美国 US447921 号专利权，如图 5-5、图 5-6 所示。

序号	专利号	专利名称	
21	US433,700	交流电电磁发电机	Alternating-Current Electro-Magnetic Motor
22	US433,701	交流电动机	Alternating-Current Motor
23	US433,702	电力传输和输入设备	Electrical Transformer Or Induction Device
24	US433,703	交流电动机	Electro-Magnetic Motor
25	US445,207	交流电动机	Electro-Magnetic Motor
26	US447,921	交流电动机	Alternating Electric Current Generator
27	US454,622	电力照明系统	System Of Electric Lighting
28	US455,067	电磁发动机	Electro-Magnetic Motor
29	US455,068	电表	Electrical Meter
30	US459,772	电磁发电机	Electro-Magnetic Motor
31	US462,418	电流转换和分配方法及装置	Method Of And Apparatus For Electrical Conversion And Distribution
32	US464,666	电磁发电机	Electro-Magnetic Motor
33	US487,796	电力传输系统	System Of Electrical Transmission Of Power
34	US511,559	电力传输系统	Electrical Transmission Of Power
35	US511,560	电力传输系统	System Of Electrical Power Transmission
36	US511,915	电力传输系统	Electrical Transmission Of Power
37	US511,916	发电机	Electric Generator
38	US512,340	电磁铁线圈	Coil For Electro-magnets
39	US512,340	电磁铁线圈	Coil For Electro-Magnets
40	US514,168	电流产生装置	Means For Generating Electric Currents

图 5-4

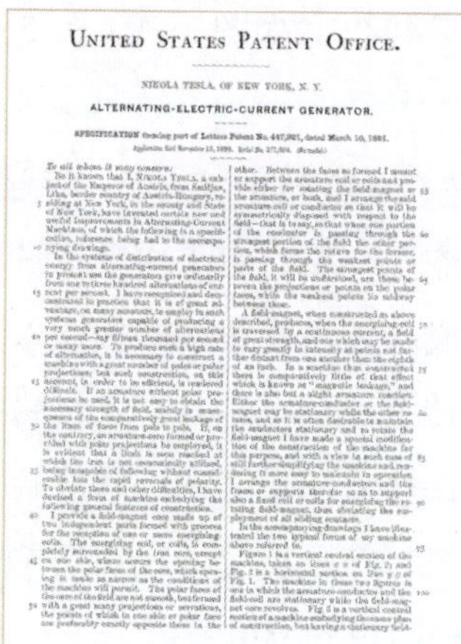

UNITED STATES PATENT OFFICE.

NIKOLA TESLA, OF NEW YORK, N. Y.

ALTERNATING-ELECTRIC-CURRENT GENERATOR.

SPECIFICATION forming part of Letters Patent No. 447,921, dated March 10, 1891.

图 5-5

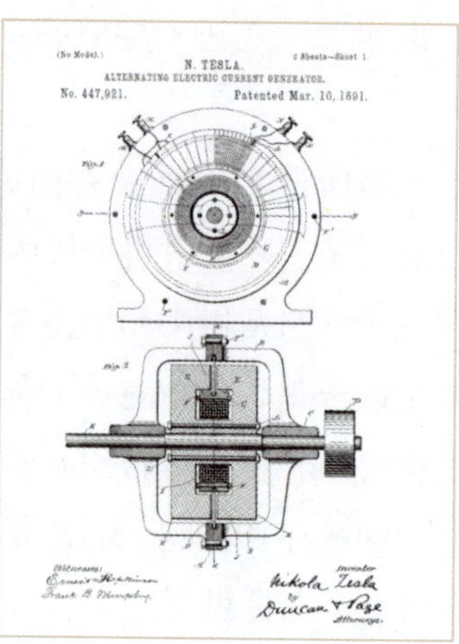

图 5-6

随着特斯拉在交流电领域研究成果的不断呈现，爱迪生与特斯拉之间关于直流电和交流电之间的竞争愈演愈烈。直到特斯拉发明了异步电动机，交流电远距离高压传输的优点也就体现出来，同时也解决了机器不能用交流电的问题。1891 年之后，多相交流发电机被广泛用于供应电流。

1895 年，特斯拉开始组建举世闻名的尼亚加拉水电站。1897 年，第一座 10 万匹马力的发电站建成，成为 35 公里外的纽约州水牛城的主要供电来源。其后 10 多座大大小小的发电站相继建成，每日所产生的电力足以供应美国纽约州和加拿大安大略省用电总需求的 1/4。至今，这项建成足足超过 100 年的电力传输系统仍然运作如常，从未间断地在产出能源，可谓是人类近百年科学史上的一大奇迹。这个科学上的百年奇迹，就是天才科学家特斯拉的设计成果，其中共运用了特斯拉的 9 项专利发明，包括他所发明的交流电发电机和交流电输电技术。

按照当时的专利权许可使用费的标准，特斯拉所拥有的交流电相关专利权，使用者每生产一匹交流电就必须向特斯拉缴纳 1 美元的专利使用费。在强大的利益驱动下，当时一股财团势力要挟特斯拉放弃相关专利权，并意图独占牟利。经过多番交涉后，特斯拉决定放弃交

流电的专利权，不再收取任何专利使用费，共享给社会。

特斯拉早在一百多年前就利用专利制度，构建了自己在交流电领域的独立王国。虽然他因主动放弃部分专利权而没有获得多少经济收益，但是对于今天的发明者和企业经营者来说他仍然可以称得上是专利战略的先行者，为当今企业专利战略的实施提供了很好的参考和借鉴。从技术领域来看，特斯拉拥有的60余项专利，覆盖了整个交流电系统，其他人想要开设交流发电装置或建造电站必须获得他的专利授权。从专利保护地域范围来看，除美国之外，特斯拉还在加拿大、英国和奥地利申请了专利，在当时的发达国家都构建了自己的专利版图。如果特斯拉当年没有放弃他的专利权而是建立自己的商业王国，他的前景是不可估量的。

特斯拉一生独自开发并取得专利700多项，合作开发达1000项以上，其中包括：1898年奠定无线电遥控技术基础的美国US613809号专利等20多项无线电领域的专利，以及1899年的X光摄影技术专利等。但特斯拉从不在意自己的财务状况，专利使用费分文未放入自己囊中。1943年1月7日在穷困且被遗忘的情况下，特斯拉在纽约人旅馆因心脏衰竭孤独地死去，享年87岁。

1960年在法国巴黎召开的国际计量大会上，电磁感应强度的单位被命名为特斯拉，简称“特”，以纪念这位伟大的发明家在电磁学领域作出的杰出贡献。

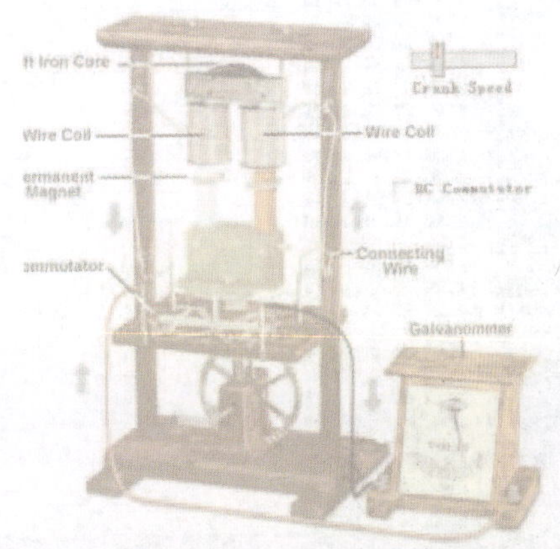

美国 America

1888.US 381968

电磁电机

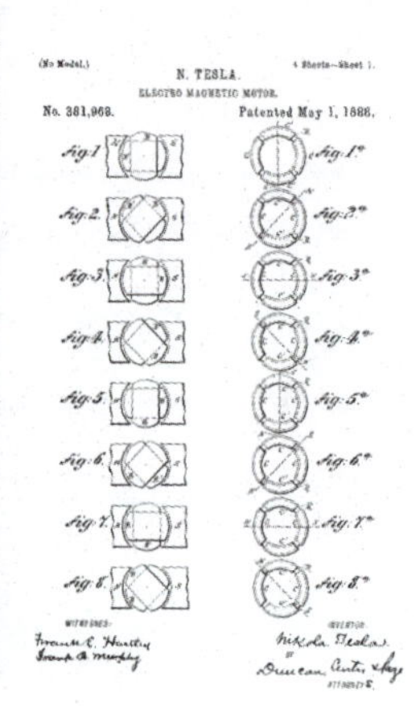

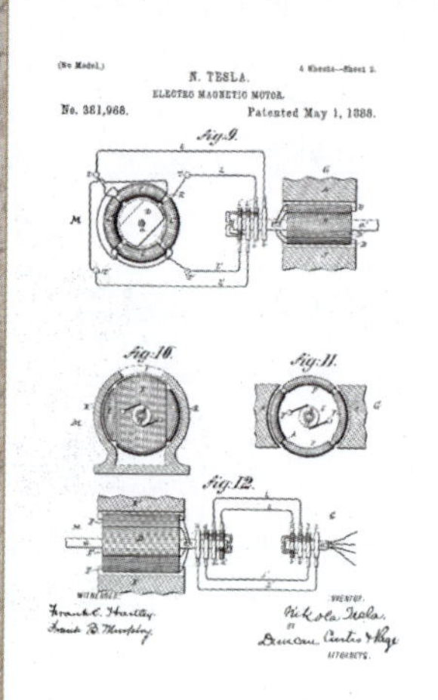

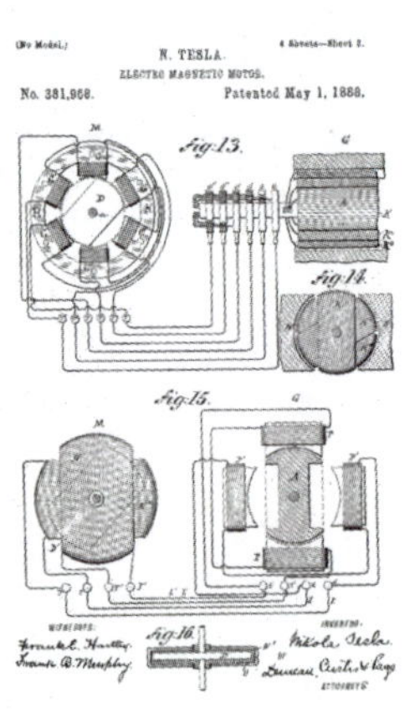

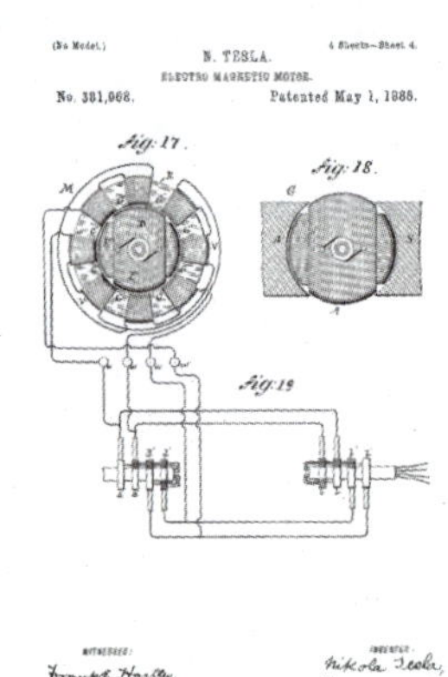

美国 America

UNITED STATES PATENT OFFICE.

NIKOLA TESLA, OF NEW YORK, N. Y., ASSIGNOR OF ONE-HALF TO CHARLES F. PECK, OF ENGLEWOOD, NEW JERSEY.

ELECTRO-MAGNETIC MOTOR.

SPECIFICATION forming part of Letters Patent No. 381,968, dated May 1, 1888.

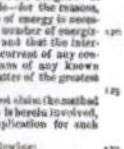

NIKOLA TESLA.

伟大发明家莱特兄弟

威尔伯·莱特（Wilbur Wright，1867–1912年）和奥维尔·莱特（Orville Wright，1871–1948年）分别出生在美国的印地安那州和俄亥俄州。他们对飞行的兴趣可以追溯至1878年父亲给他们的一个飞行玩具。

1903年，莱特兄弟制造出了世界上第一架依靠自身动力进行载人飞行的飞机——“飞行者1号”，12月试飞成功。1906年，他们的飞机在美国获得专利权。

1909年，莱特兄弟获得美国国会荣誉奖。同年，他们创办了莱特飞机公司。

US 821393

www.sipo.gov.cn

滑翔机的发明

Glider

奥托·李林塔尔

Otto Lilienthal

人类很早就憧憬像鸟儿一样在天空飞翔，而且从未停止过这样的探索。15世纪伟大的艺术家达·芬奇就曾设计过扑翼机。19世纪初，乔治·凯利爵士提出了空气动力学理论之后，奥托·李林塔尔于1891年成功制造了第一架实用滑翔机。随后莱特兄弟发明了第一架动力飞机，开辟了人类航空史的新篇章。

奥托·李林塔尔，1848年出生在德国。1891年，他制成一架蝙蝠状的拱形翼滑翔机，成功地进行了滑翔飞行。在1891～1896年间，他先后制造了18种不同的滑翔机，并于1893年获得一项德国专利（DE77916）。直到1896年在一次试飞中不幸丧命，他前后共进行了2000多次滑翔试验，积累了大量数据，并以此编制了《空气压力数据表》，给后来的飞机制造者们提供了宝贵的资料。

奥托·李林塔尔最早设计和制造出实用的滑翔机，被尊称为德国的“滑翔机之父”。

DE 77916

第六篇 飞 机

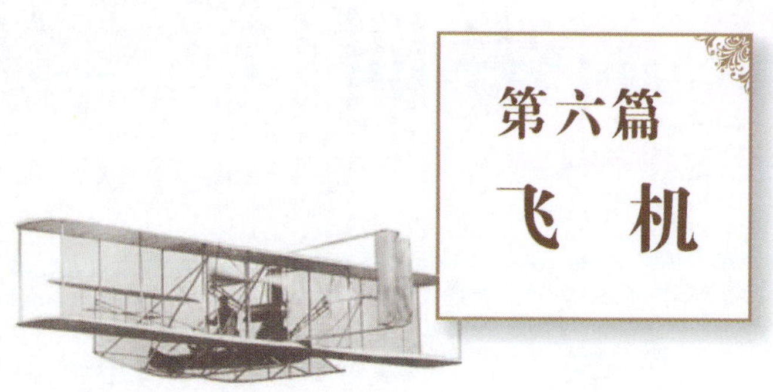

人类自古以来就梦想着能像鸟儿一样在天空中自由地飞翔。2000多年前中国人发明的风筝，虽然不能将人带上天空，但它确实可以被称为飞机的鼻祖。19世纪末许多国家都掀起了飞机研制与试飞的热潮。这些试飞活动虽然都曾离开过地面一段距离，但还不是现代意义上的“飞”，或者说它们只是靠一定动力推动的跳跃而已。

威尔伯·莱特（Wilbur Wright，1867-1912年）和奥维尔·莱特（Orville Wright，1871-1948年）分别出生在印第安那州和俄亥俄州。兄弟二人从小对飞行十分感兴趣，他们通过观察鸟的飞行，想象着人类有一天也可以飞向天空。1893年德国滑翔机先驱奥托·李林塔尔（Otto Lilienthal）发明并获得名为“滑翔机”的德国DE77916号专利权，及试飞成功的消息使莱特兄弟立志飞行。1896年奥托·李林塔尔试飞失事，促使莱特兄弟把注意力集中在滑翔机的平衡操纵上面，并决定仿制一架滑翔机。他们观察老鹰飞行的动作，并学习了大量航空方面的知识，经过反复试验，终于在1900年研制出第一架“自动飞行机”。这个“自动飞行机”的外形类似老鹰展翅，有两个机翼，机翼下面系一条长绳，放飞时需像放风筝一样助跑，从而借助风力起飞。最初，这架 “自动飞行机”只能飞到1米多高，经过半年多的改良后，可以飞到100多米的高度。然而助跑速度与绳子长度是有限的，要解决滑翔机操纵这个关键问题，必须有某种能发挥作用的装置。按照这一想法，经过相当长时间的试验之后，莱特兄弟终于设计制造出了人工操作性更好也更安全的载人滑翔机，并于1903年申请

了名为“飞机器”的美国 US821393 号专利权，如图 6-1 所示。

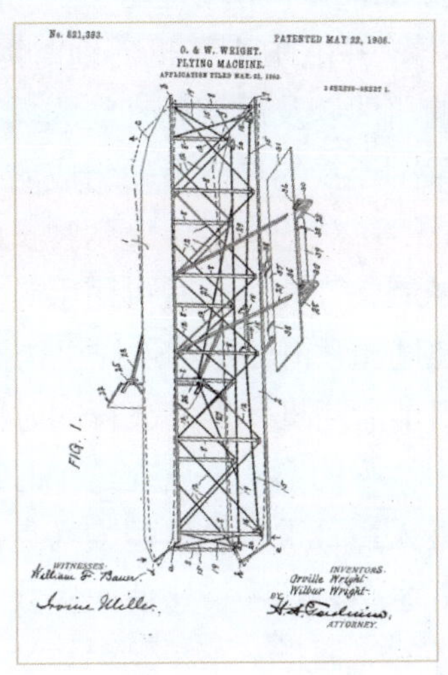

图 6-1

从该专利说明书的内容来看，虽然这项申请的名称为“飞机”，但并未包括动力装置和螺旋桨，仍是一种滑翔机。其主要内容包括以下几点：（1）飞机包括机翼、尾部垂直舵和前部水平舵；机翼从边端到边端的宽度显著地大于飞机从前端到尾端的长度；飞机的结构件是由轻质而牢固的木材之类的材料制成的，机翼的框架上紧绷着布之类的高强度纤维织物。（2）机翼或是单机翼的或是多机翼的，说明书和附图所列举的例子是平行、垂直重叠、由一系列平行柱相连接间隔一定距离的双机翼。所说的机翼是扁平的，具有水平边缘部分，可通过操作使该水平边缘部分与飞行方向垂直的机翼平面的横轴扭曲，并与常态的机翼形成不同角度，使之与气流之间形成升力和平衡。（3）尾部垂直舵由 V 形支架与上下机翼相连接，其主要作用是可围绕垂直枢轴转动，从而形成对空气的不同阻力，使飞机可绕纵轴改变左右方向和平衡，该尾部垂直舵也可绕 V 形支架和上、下机翼连接的水平枢轴绕动，以致尾部垂直舵上升，这样可防止飞机降落着地时该垂直舵与地面摩擦碰撞损坏。（4）前部水平舵安装在由下机翼中部左右各一根水平支撑杆和上机翼中部相应的左右各一根支撑斜杆组成的架子上，上述水平支撑杆前端向上凸起端头。前部水平舵通过水平枢轴与穿过它的斜杆相连，该水平舵通过弹簧与水平支撑杆前

端向上凸起的端头相连，使该水平舵一般处于水平状态，也使其具有柔性；通过操作水平舵可绕水平枢轴绕动一定角度，从而改变空气阻力，使飞机向上和向下，并能防止飞机着地时颠簸、倾翻。（5）飞机机翼水平边缘部分的扭曲以及尾部垂直舵的操作系统是由拉紧的绳索和一组与机翼、机架相连接的导向滑轮组成的，绳索两端分别牢固系于作为方便操作装置、吊架的侧边上，驾驶员躺在吊架上，用身体运动对飞机进行操作控制，驾驶员同时可用手等方式操作控制前部水平舵，该水平舵也是通过绳索滑轮组和连杆机构操作的。

虽然成功地研制出了滑翔机，但是飞行的时间太短，不能满足莱特兄弟想长时间在空中飞行的愿望。就在这个时期，航空事业连连受挫，飞行技师皮尔机毁人亡，重机枪发明人马克沁试飞失败，航空学家兰利连飞机带人掉入水中，等等，这使大多数人认为飞机依靠自身动力飞行是完全不可能的。但是，莱特兄弟却没有放弃自己的努力。他们把有关飞行的资料集中起来，反复研究，却始终想不到用什么动力能将庞大的滑翔机和人运到空中。突然有一天，哥哥威尔伯·莱特想到，汽车有发动机就可以开动，那为什么不在滑翔机上面装一个发动机呢？假如给滑翔机加装动力并带上足够的燃料，那么它是不是就可以自由地飞翔、起降呢。于是，莱特兄弟俩积极努力实践自己的大胆想法，开始了动力飞机的研制。经过反复试验，他们研制出了内燃发动机和螺旋桨，并且将自己制造的带螺旋桨和发动机的飞机模型，放到自制的“风洞”中去模拟飞行。1903 年 9 月，他们将自己制造的“飞行者 1 号”飞机，一架 12 马力、四缸活塞式发动机驱动的飞机，拉到东海岸的基蒂·霍克海滩，进行了充分的试飞准备。1903 年 12 月 17 日，这是一个不寻常的日子，弟弟奥维尔·莱特和哥哥威尔伯·莱特分别驾驶着“飞行者 1 号”飞机，成功地飞行了四次，总共飞了 97 秒，飞了 441 米远。虽然只有短短的时间和距离，但这却是人类第一次真正地乘动力飞机飞行。莱特兄弟终于发明了动力可操控飞机，人类就此进入飞行时代。

值得一提的是，关于飞机的发明权属问题还有一段插曲。1901 年，美国政府曾出资 5 万美元，支持当时著名的研究者兰利（Langley，Samuel Pierpont）进行飞机研究。1903 年，莱特兄弟成功飞行的前 70 天，兰利研制的“航空站号”飞机也进行试飞，可

惜失败了。莱特兄弟成功后，美国另一位飞行家格伦·哈蒙德·寇蒂斯（Glenn Hammond Curtis）对“航空站号”进行改装，并重新试飞成功。于是，他宣称，第一架飞机的发明者不是莱特兄弟，而是兰利。由于兰利是当时美国地位显赫的史密森学会（Smithsonian Institution）的主席，他竟滥用职权，宣布“航空站号”飞机是最早成功的飞机。直到 1942 年，新任主席才纠正了过去的错误声明，为莱特兄弟正名。同时决定，将莱特兄弟的“飞行者 1 号”陈列在美国博物馆的最佳位置。

莱特兄弟的“飞行者 1 号”试飞成功之后，各种飞机先后出现，1909 年，法国人路易·布雷里奥的单翼机第一次飞越英吉利海峡；1910 年，罗马尼亚人亨利·科安达驾驶他自己设计的喷气式飞机出现在法国巴黎展览会上；1919 年，第一架客机诞生。20 世纪 30 年代后期，活塞驱动的螺旋桨飞机的最大平飞时速已达到 700 公里，俯冲时已接近音速。此外，为节省起飞空间，1939 年 9 月 14 日世界上第一架实用型直升机诞生，是现代直升机的鼻祖。

1947 年喷气式飞机实现了超音速飞行。后来，飞机逐渐在民用、军用及航天等领域被广泛应用，性能各异，性价比越来越优化。以民用飞机为例，欧洲空客的空中巨无霸 A380，双层通道，起飞重量达 421 吨，载客量 555 人，最大载客量可达 840 人，其升空 2 年后，波音公司新一代“梦幻飞机”波音 787 下线，这是有史以来第一次机体结构的一半左右是用更轻、更坚固的碳纤维材料制作的民用飞机，它可一次加油持续飞行 1.575 万公里，载客量可达 330 人。

很快，人类已经不满足于“飞上天”，研究者们向太空探索的热情也持续高涨。1972 年，美国开始研制航天飞机，经过 10 年努力，耗费了上百亿美元的资金，投入约 5 万名科学家和工程技术人员，1981 年，搭载 2 名宇航员的航天飞机“哥伦比亚”号成功冲上太空，在绕行地球 36 圈后安全着陆，翻开了人类航天史上崭新的一页。

1906.US 821393

——飞行器

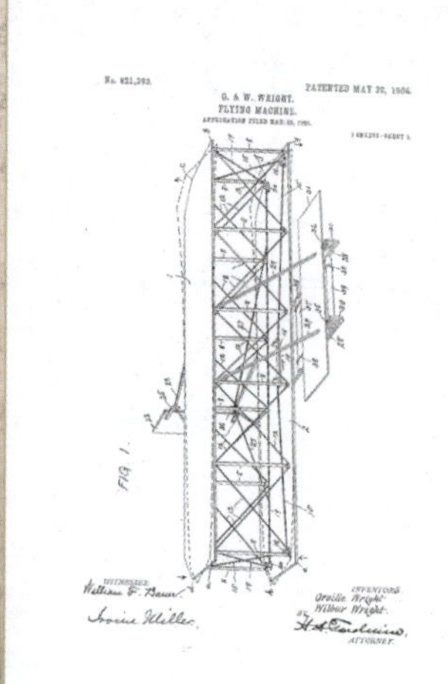

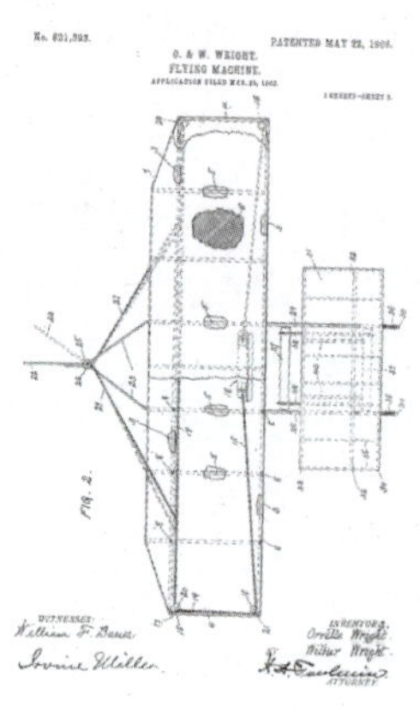

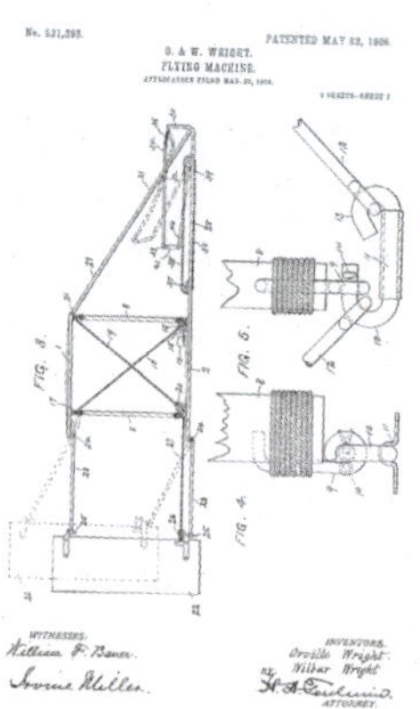

UNITED STATES PATENT OFFICE.

ORVILLE WRIGHT AND WILBUR WRIGHT, OF DAYTON, OHIO.

FLYING-MACHINE.

美国
America

德国 German

1893. DE 77916

飞行器

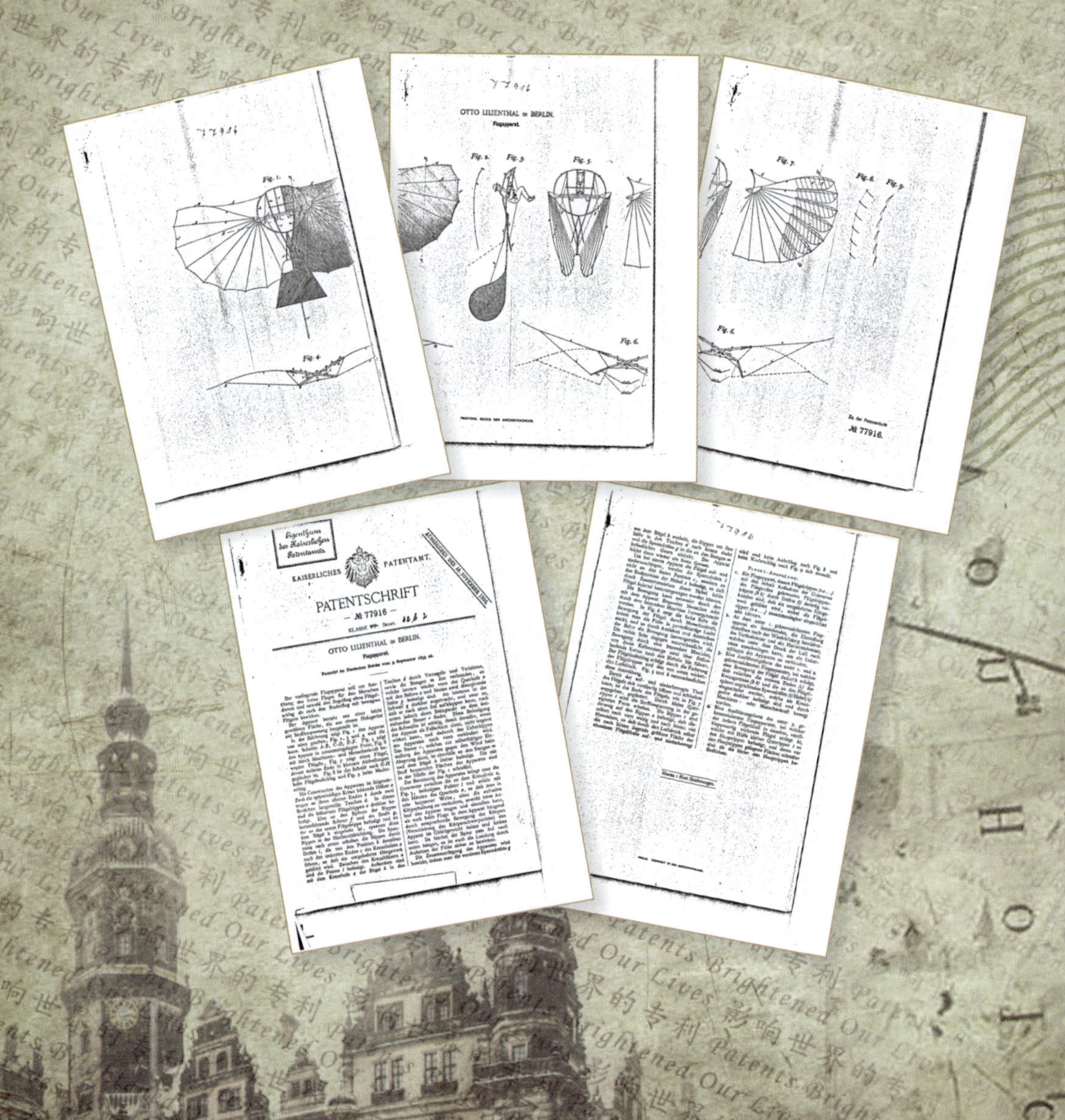

安全剃须刀的发明 RAZOR

金·坎普·吉利
King Camp Gillette

安全剃须刀的出现，是日常生活用品之树上绽开的一朵绚丽多彩的奇葩。在目前的市场上仍能见到性能各异、琳琅满目的安全剃须刀产品。而它的发明者就是美国人金·坎普·吉利。

吉利感到当时的直刃剃须刀需经常打磨刃口，外出时携带不便，而且剃须时一不小心就会划破脸皮，很不安全。他初步设想在两块薄板之间夹上两边有刃口的钢制刀片，用一个螺栓手柄固定组合起来组成新型的剃须刀代替直刃剃须刀。整整花了 6 年时间，研制成功具有硬度、锋利、两边刃缘的廉价刀片，解决了刀片刃口的保护和刀架边缘之间的精确调节定位，确保剃须时不伤及皮肤。该项发明于 1904 年被授予美国专利 US 775134。

申请专利后，吉利努力实施，开拓市场，使发明产业化，在商业上获得巨大成功，创造了吉利剃须刀百年不衰的奇迹。吉利创意性地将自己的脸印刷在产品包装上，也成为吉利剃须刀产品的一大标志性特色。

US 775134

www.sipo.gov.cn

第七篇 安全剃须刀

剃须刀是男士们必备的日常用品。在20世纪，吉利剃须刀已经成为世界知名的品牌，吉利正是安全剃须刀的发明人，那张印在剃须刀包装上的吉利的画像，随着吉列剃须刀销往全世界而成为为人所熟知的面孔。

剃须刀的发明人金·坎普·吉利（King Camp Gillette，1855-1932年），美国人，是芝加哥的一名推销员。19世纪末期的美国男士们使用的剃须刀既笨重，又不锋利，刮脸费时费力，稍不留神还会刮破脸。由于刀身不能更换，要使剃须刀好使一些，只有频繁地磨刀。当时磨刀有两种办法：一是送到专业磨刀店里去研磨，费时又费钱；二是在刀布上来回磨。吉利也常常因为刮须刀不顺手而刮破脸，他工作之余常思考如果能有一种轻便、锋利、安全的剃须刀来代替这种老式剃须刀，肯定有销路。吉利买来锉刀、夹钳、薄钢片等工具和材料，专心地研究起来。他想，代替刀身的薄刀片可以"用完即扔"，但刀片必须能和刀柄分开。于是，吉利把刀柄设计成圆形，上方留有凹槽，从而能用螺丝把刀片固定。刀片用超薄型钢片制作，并夹在两块薄金属片中间，露出刀刃，使用时刀刃与脸部始终可形成固定的角度。这样，既能方便地刮掉脸部和下巴上任何部位的胡须，又不容易刮破脸。确定设计方案后，吉利请专业人员制作出样品，但是使用效果不是很理想。

1901 年，吉利认识了麻省理工学院毕业的机械工程师尼克逊（Nickerson），并与他沟通了自己的想法。尼克逊对此很感兴趣。在尼克逊的帮助下，经过反复试验，吉利的安全剃须刀终于可以投入使用，1901 年吉利将他的发明在美国提交了专利申请，并于 1904 年获得名为“剃须刀（Razon）”的美国 US775134 号专利权，如图 7–1、图 7–2 所示。尼克逊成为吉利的合伙人，共同成立了美国安全剃须刀公司（后来改名为吉利安全剃须刀公司）。

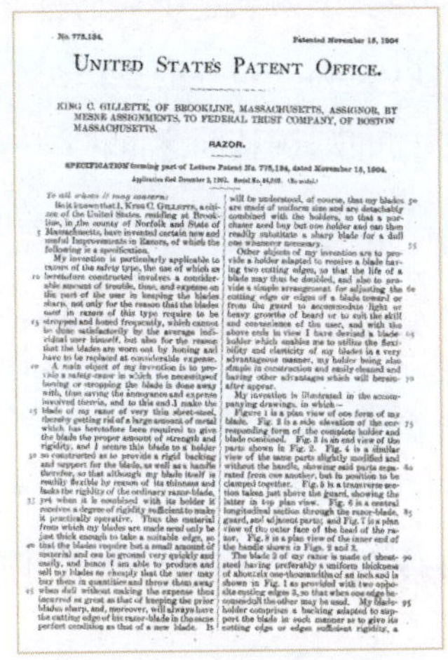

No. 775,134. Patented November 15, 1904

UNITED STATES PATENT OFFICE.

KING C. GILLETTE, OF BROOKLINE, MASSACHUSETTS, ASSIGNOR, BY MESNE ASSIGNMENTS, TO FEDERAL TRUST COMPANY, OF BOSTON MASSACHUSETTS.

RAZOR.

SPECIFICATION forming part of Letters Patent No. 775,134, dated November 15, 1904.

To all whom it may concern:

Be it known that I, KING C. GILLETTE, a citizen of the United States, residing at Brookline, in the county of Norfolk and State of Massachusetts, have invented certain new and useful Improvements in Razors, of which the following is a specification.

图 7–1

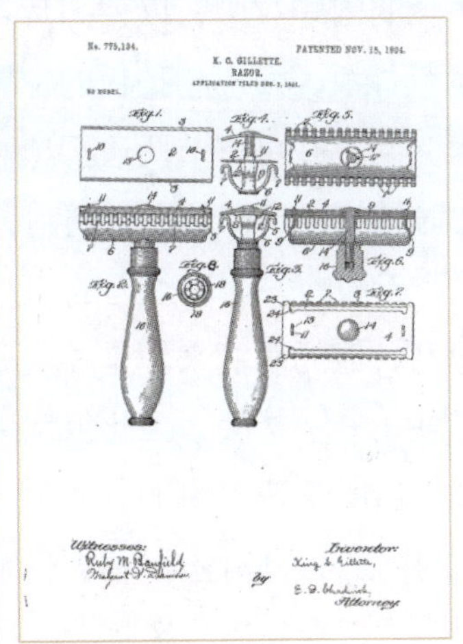

图 7–2

在这项发明中，吉利将剃须刀分为两部分，剃须刀刀片和刀架，刀片可从刀架上拆卸下来。刀片由 3%～6% 英寸均匀厚度、可在横向上弯曲的挠性钢片制成，其两边各为淬过火的刃口，当一边刃口钝了就可使用另一边的刃口。刀架包括背板、护板和手柄；背板为具有足够刚度的金属板，安在刀架顶端，它与刀片同长，但比刀片稍窄；护板由金属片冲压成近似半柱面型，在其中心形成孔，面向背板形成中凹形状，便于剃须时贮存肥皂泡沫，背板两边的边缘各设有外伸拱起的一排齿；护板和背板将刀片夹在当中，当安装刀片后，可限制刀片刃口暴露，避免伤害使用者皮肤；手柄安装于刀架的中心位置，并在安装时把背板、护板和刀片安装组合在一起，同时能调节刃口的位

置。背板、护板和刀片的安装是利用顶部的螺栓固定在手柄上。刀片的两个刃口相对于刀架对称，以便两边刃口都能同样使用。在护板内，安装在手柄和背板之间的中心位置有一中空柱，中空柱上部形成纵向延伸的窄形分臂支撑板，支撑板和背板具有一致的弧形，将刀片可靠地夹住。护板两侧设有支柱，支柱延伸出背板的部分形成自由端。上述这些部件是经过如下装配并形成对刃口位置的调节：螺栓穿过相应背板的中心孔、刀片的中心孔、支撑板的中心孔和中空柱的中心孔，紧固在手柄的内螺纹中，以便刀片可靠装于刀架之中。通过螺栓的作用，可调节背板、刀片和护板的位置以及确保刀片的刃口可夹持在托板和护板之间；并可调节两边刃口和齿的圆拱基面之间的位置。本发明在使用时握持手柄即可进行剃刮，并不会伤及皮肤，安全剃须。

其实，吉利的安全剃须刀的生产成本很高。但是吉利并不靠出售剃须刀赢利，他的利润来源是 5 美分一片的吉利专利刀片。由于一个刀片可以使用 6 ~ 7 次，因此每刮一次脸所花的钱不足 1 美分，甚至不到去理发店所花费用的 1/10。剃刀和刀片分离，新刀片瞬间就可安装上，省时省力，刮时不但不会伤及皮肤，而且舒适无比。所以吉利剃须刀上市后，受到热烈欢迎。

吉利在发明了安全剃须刀之后仍继续研究，不断改进方案升级产品。从专利文献中可以查到，吉利的发明生涯中一共申请了 11 项专利。这些发明一部分是围绕剃须刀的改进和辅助产品，如 1903 年获得名为“剃须刀安全性的改进”的英国 GB28763 号专利权，如图 7-3、图 7-4 所示，还有 1915 年获得名为“刀片盒”的美国 US1132925 号专利权；另一部分是吉利在其推销员生涯中发明的相关产品，如 1905 年获得名为“密封瓶盖的改进”的英国 GB13232 号专利权。

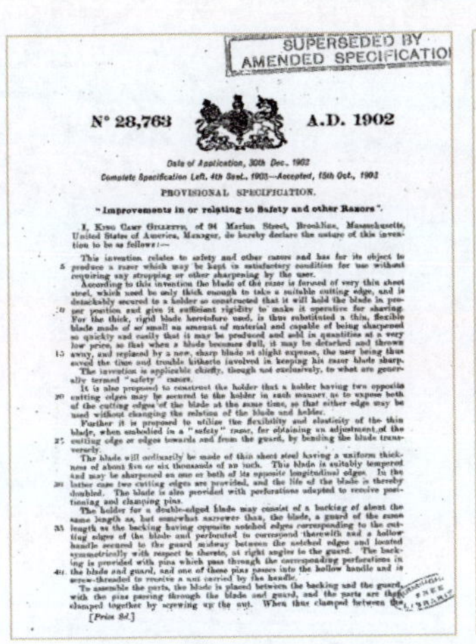

SUPERSEDED BY AMENDED SPECIFICATION

Nº 28,763　　A.D. 1902

Date of Application, 30th Dec., 1902

Complete Specification Left, 4th Sept., 1903—Accepted, 15th Oct., 1903

PROVISIONAL SPECIFICATION.

"Improvements in or relating to Safety and other Razors".

I, King Camp Gillette, of 94 Marion Street, Brookline, Massachusetts, United States of America, Manager, do hereby declare the nature of this invention to be as follows:—

This invention relates to safety and other razors and has for its object to produce a razor which may be kept in satisfactory condition for use without requiring any stropping or other sharpening by the user.

According to this invention the blade of the razor is formed of very thin sheet steel, which need be only thick enough to take a suitable cutting edge, and is detachably secured to a holder so constructed that it will hold the blade in proper position and give it sufficient rigidity to make it operative for shaving. For the thick, rigid blade heretofore used, is thus substituted a thin, flexible blade made of so small an amount of material and capable of being sharpened so quickly and easily that it may be produced and sold in quantities at a very low price, so that when a blade becomes dull, it may be detached and thrown away, and replaced by a new, sharp blade at slight expense, the user being thus saved the time and trouble hitherto involved in keeping his razor blade sharp.

The invention is applicable chiefly, though not exclusively, to what are generally termed "safety" razors.

It is also proposed to construct the holder that a holder having two opposite cutting edges may be secured to the holder in such manner as to expose both of the cutting edges of the blade at the same time, so that either edge may be used without changing the relation of the blade and holder.

Further it is proposed to utilise the flexibility and elasticity of the thin blade, when embodied in a "safety" razor, for obtaining an adjustment of the cutting edge or edges towards and from the guard, by bending the blade transversely.

The blade will ordinarily be made of thin sheet steel having a uniform thickness of about five or six thousands of an inch. This blade is suitably tempered and may be sharpened on one or both of its opposite longitudinal edges. In the latter case two cutting edges are provided, and the life of the blade is thereby doubled. The blade is also provided with perforations adapted to receive positioning and clamping pins.

The holder for a double-edged blade may consist of a backing of about the same length as, but somewhat narrower than, the blade, a guard of the same length as the backing having opposite notched edges corresponding to the cutting edges of the blade and perforated to correspond therewith and a hollow handle secured to the guard midway between the notched edges and located symmetrically with respect to thereto, at right angles to the guard. The backing is provided with pins which pass through the corresponding perforations in the blade and guard, and one of these pins passes into the hollow handle and is screw-threaded to receive a nut carried by the handle.

To assemble the parts, the blade is placed between the backing and the guard, with the pins passing through the blade and guard, and the parts are then clamped together by screwing up the nut. When thus clamped between the

[Price 8d.]

图 7–3

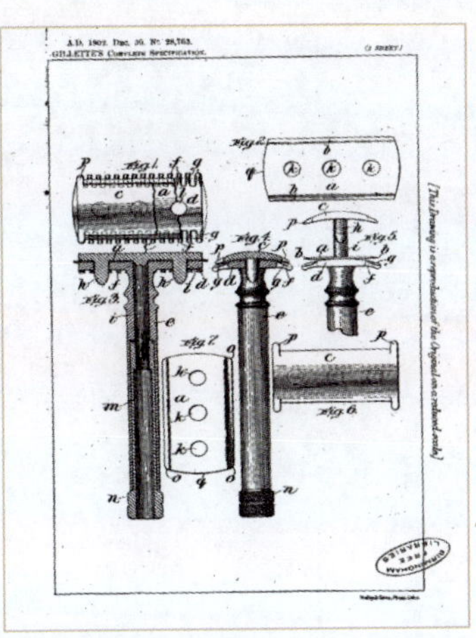

图 7–4

直到今天，剃须刀家族仍然在不断地推陈出新，刀片更薄、更坚固耐用；20 世纪中期出现了电动剃须刀，更是大大节省了男士们打理胡须的时间。吉利公司也在产品上不断创新，它已经不单单生产剃须刀，还增加了个人家庭用品系列。但作为吉利最重要的产品，剃须刀带来的利润是无法估算的。无法想象，没有了这个看似不起眼的东西，我们的生活会变成什么样，所以应该感谢吉利——这个创始人最初那一刹那的灵感，给全人类带来了便捷和舒适！

美国 America

1904.US 775134

剃须刀

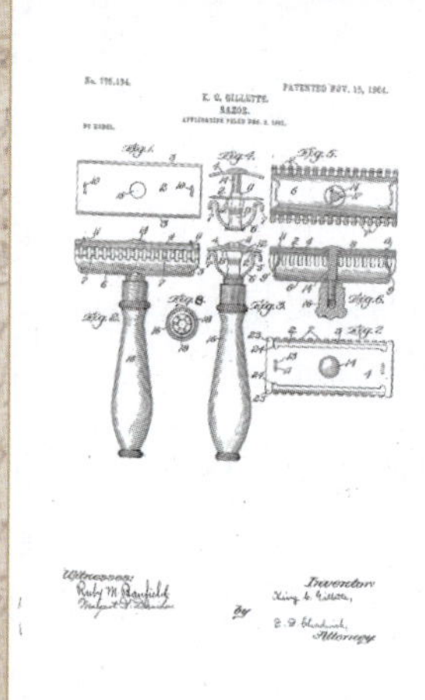

吸尘器的发明 Vacuum Cleaner

胡伯特·布斯 Hubert Booth

1900 年，英国工程师胡伯特·布斯在一个贸易展上参观美国一种车箱除尘器示范表演，这种机器用压缩空气把尘埃吹入容器内，布斯在此基础上发明了世界上第一台吸尘器。

经多次实验，布斯制成了吸尘器，用强力电泵把空气吸入软管，通过布袋将灰尘过滤。1901 年 8 月布斯取得英国专利 GB17433，并成立了真空吸尘公司。公司并不出售吸尘器，而是提供吸尘服务。吸尘器的真空状态由一个吸气泵产生，后来又以多层涡轮风扇来取代。

1906 年布斯制成了家庭小型吸尘器，虽名为“小型”，但却重达 88 磅（1 磅＝0.4536 千克），因太笨重而无法普及。后来美国发明家斯班格拉制成了更加轻巧的吸尘器，成为最早的家用吸尘器设计，其原理一直使用至今。

GB 17433

www.sipo.gov.cn

第八篇 吸尘器

吸尘器是现代生活中最常见的电器之一。今天我们所使用的吸尘器的全称是真空吸尘器，它的基本原理就是利用空气泵（后被涡轮风扇取代）在密封的容器内创造一个局部的真空环境，利用空气负压将灰尘和污垢吸入。在 20 世纪初，英国工程师胡伯特·布斯（Hubert Booth，1871–1955 年）利用这一原理发明了世界上第一台电动的真空吸尘器。此后，现代的各类吸尘器都是在真空吸尘这一原理上发展而来。而在真空吸尘器出现以前，除尘的方式有的是将地毯挂起来反复拍打，也有的是使用风箱鼓风吹出灰尘。直到电动真空吸尘方式的出现，吸尘器才开始在人类的生活中得以广泛使用，成为实用的清洁工具。

1900 年，布斯在一个贸易展上看到一种来自美国的车厢除尘器，它的除尘方式是用压缩空气将灰尘吹入容器内。布斯认为这种方式的除尘效果并不理想，但给他的启发是反其道而行之——吸尘。他做了一个简单的试验：将一块手帕搭在椅子扶手上，用口对着手帕吸气，结果手帕附上了一层灰尘。基于这种原理，他用强力电泵把空气吸入软管，通过布袋将灰尘过滤，制成了真空吸尘器。1901 年，布斯获得了名为“涉及地毯及其他材料灰尘提取的改进”英国 GB17433 号专利权，如图 8–1、图 8–2 所示。

通过这篇专利文献我们了解到，布斯发明的真空吸尘器如图 8–2

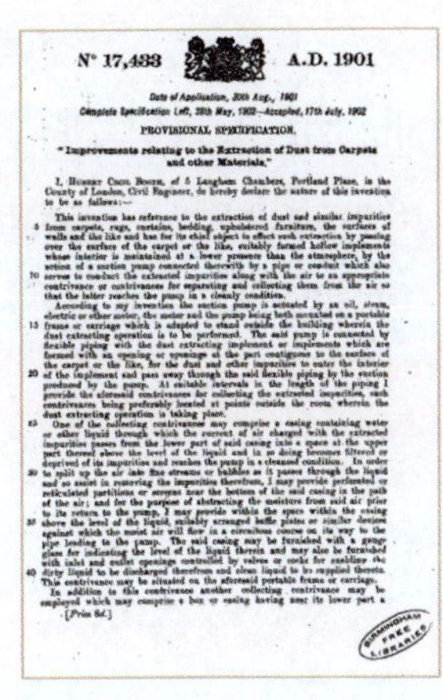

Nº 17,433 A.D. 1901

Date of Application, 30th Aug., 1901

Complete Specification Left, 28th May, 1902—Accepted, 17th July, 1902

PROVISIONAL SPECIFICATION.

"Improvements relating to the Extraction of Dust from Carpets and other Materials."

I, HUBERT CECIL BOOTH, of 5 Langham Chambers, Portland Place, in the County of London, Civil Engineer, do hereby declare the nature of this invention to be as follows:—

This invention has reference to the extraction of dust and similar impurities from carpets, rugs, curtains, bedding, upholstered furniture, the surfaces of walls and the like and has for its chief object to effect such extraction by passing over the surface of the carpet or the like, suitably formed hollow implements whose interior is maintained at a lower pressure than the atmosphere, by the action of a suction pump connected therewith by a pipe or conduit which also serves to conduct the extracted impurities along with the air to an appropriate contrivance or contrivances for separating and collecting them from the air so that the latter reaches the pump in a cleanly condition.

According to my invention the suction pump is actuated by an oil, steam, electric or other motor, the motor and the pump being both mounted on a portable frame or carriage which is adapted to stand outside the building wherein the dust extracting operation is to be performed. The said pump is connected by flexible piping with the dust extracting implement or implements which are formed with an opening or openings at the part contiguous to the surface of the carpet or the like, for the dust and other impurities to enter the interior of the implement and pass away through the said flexible piping by the suction produced by the pump. At suitable intervals in the length of the piping I provide the aforesaid contrivances for collecting the extracted impurities, such contrivances being preferably located at points outside the room wherein the dust extracting operation is taking place.

One of the collecting contrivances may comprise a casing containing water or other liquid through which the current of air charged with the extracted impurities passes from the lower part of said casing into a space at the upper part thereof above the level of the liquid and in so doing becomes filtered or deprived of its impurities and reaches the pump in a cleansed condition. In order to split up the air into fine streams or bubbles as it passes through the liquid and so assist in removing the impurities therefrom, I may provide perforated or reticulated partitions or screens near the bottom of the said casing in the path of the air; and for the purpose of abstracting the moisture from said air prior to its return to the pump, I may provide within the space within the casing above the level of the liquid, suitably arranged baffle plates or similar devices against which the moist air will flow in a circuitous course on its way to the pipe leading to the pump. The said casing may be furnished with a gauge glass for indicating the level of the liquid therein and may also be furnished with inlet and outlet openings controlled by valves or cocks for enabling the dirty liquid to be discharged therefrom and clean liquid to be supplied thereto. This contrivance may be situated on the aforesaid portable frame or carriage.

In addition to this contrivance another collecting contrivance may be employed which may comprise a box or casing having near its lower part a

[Price 8d.]

图 8-1

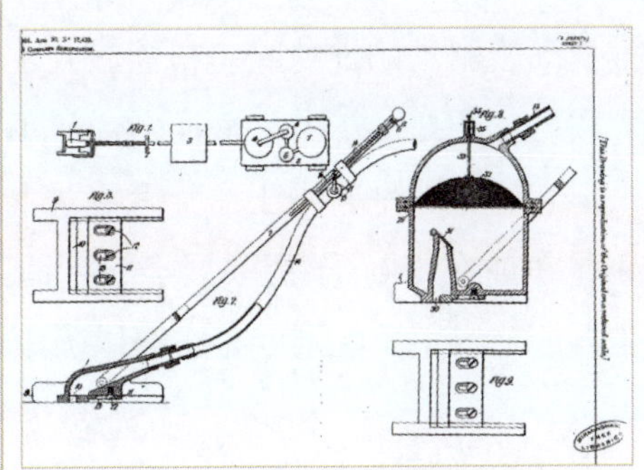

图 8-2

所示。它的最前端为一个吸尘口，与吸尘口相连的是一根用于手持的长柄，吸尘口通过软管与后端的灰尘收集容器相联。这个吸尘器设计了两个灰尘收集容器：第一个容器里设有一个圆顶形的过滤罩，进行灰尘的第一次过滤；第二个容器用于盛水，通过水进行灰尘的二次过滤。灰尘收集容器又与吸气泵连接，吸气泵由一台电机驱动。由于带有笨重的电机和吸气泵，这台吸尘器使用时需要安装在马车上移动。

随后，布斯成立了真空吸尘公司，第一次在市场上使用了“真空吸尘器”的名称。但他的公司不出售吸尘器，而是提供吸尘服务。1902 年这一吸尘器被指定为爱德华七世加冕典礼清洁地毯，后来又被指定为白金汉宫清洁，引起了轰动，人们开始追捧这种新式的清洁方式。1906 年布斯制成了家庭小型吸尘器，但仍重达近 80 斤，无法普及。在后来的发展中，他的产品逐渐主要面向工业用途。

而在大西洋另一侧的美国，发明家们也在为能研制出轻便高效的除尘设备而努力。早在 1869 年，美国人艾维斯 · W. 麦加菲（Ives W. McGaffey）就发明了一台真空吸尘器，命名为“旋风”（Whirlwind），并获得美国 US91145 号专利权，如图 8-3、图 8-4 所示。这种吸尘器的外观与我们今天使用的立式吸尘器相似，它也采用了真空吸尘的原理。但与布斯的电动真空吸尘器不同，它是手动的——使用者需要通过摇动吸尘器尾部的曲柄来带动风扇旋转，以吸入灰尘。手动吸尘器的操作非

常费力和不方便，因此没能得到普及。

1898 年，另一个美国人约翰·瑟曼（John Thurman）发明了一种靠汽油发动机驱动的地毯除尘机器，并于 1899 年获得美国 US634042 号专利权，如图 8-5、图 8-6 所示。这项发明也利用空气流动来去除灰尘，但它的原理是利用压缩空气将灰尘吹出，而不是吸尘——这很可能正是布斯在 1900 年贸易展上看到的除尘器。而布斯的发明与这种除尘器的一点主要区别就是布斯以吸尘取代了吹尘。

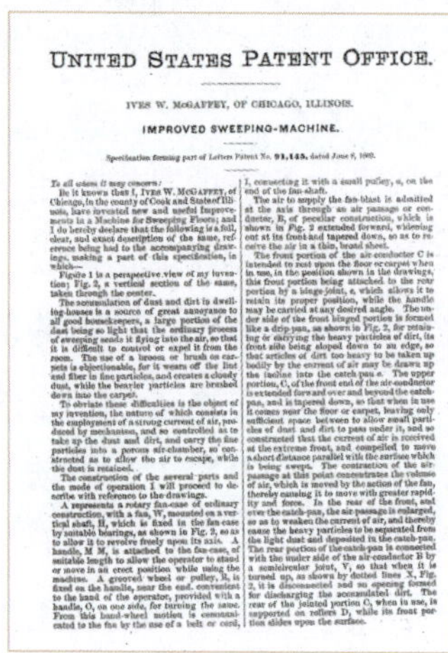

UNITED STATES PATENT OFFICE.

IVES W. McGAFFEY, OF CHICAGO, ILLINOIS.

IMPROVED SWEEPING-MACHINE.

Specification forming part of Letters Patent No. 91,145, dated June 8, 1869.

图 8-3

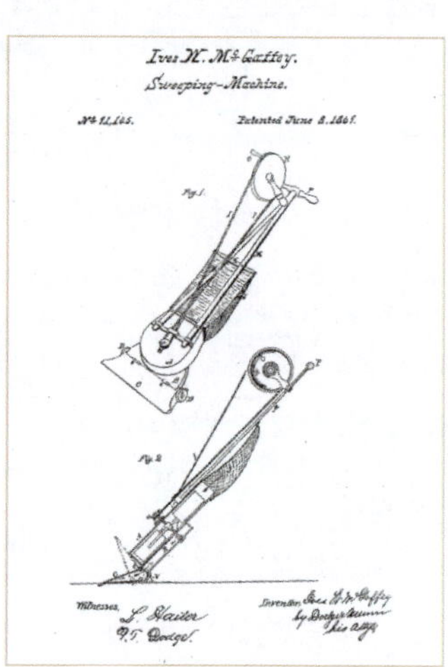

图 8-4

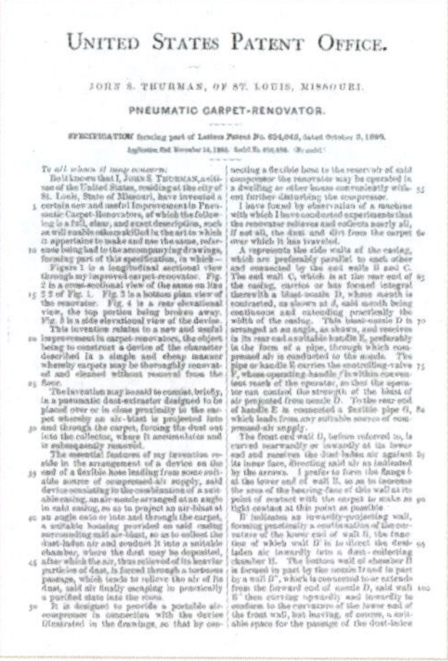

UNITED STATES PATENT OFFICE.

JOHN S. THURMAN, OF ST. LOUIS, MISSOURI.

PNEUMATIC CARPET-RENOVATOR.

图 8-5

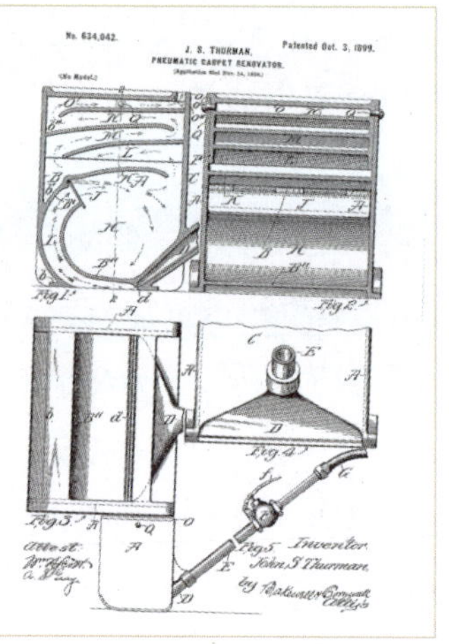

图 8-6

而现代家用吸尘器的雏形源自美国发明家詹姆斯·M. 斯班格拉（James. M. Spangler）于 1907 年发明的一种移动式吸尘器。它带有风机、旋转刷和集尘袋，比布斯的吸尘器更加轻便——这就是最早的家用吸尘器的设计。斯班格拉为他的发明在美国申请了专利，于 1908 年获得美国 US889823 号专利权，如图 8-7、图 8-8 所示。

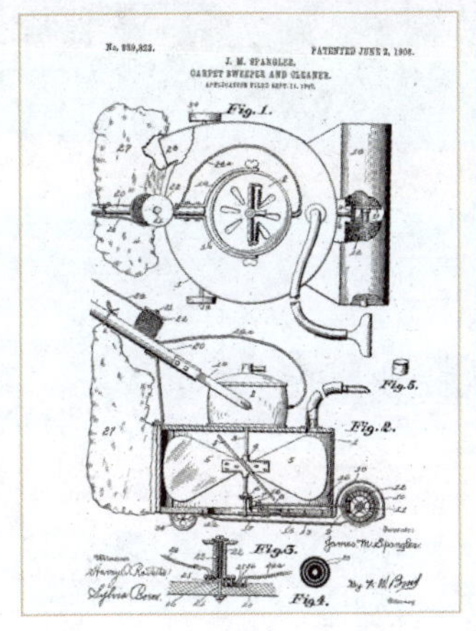

图 8-7

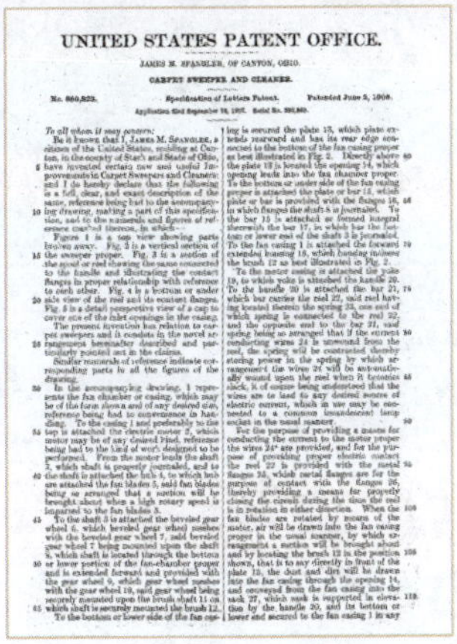

UNITED STATES PATENT OFFICE.

JAMES M. SPANGLER, OF CANTON, OHIO.

CARPET SWEEPER AND CLEANER.

图 8-8

获得授权后不久，斯班格拉将这项专利权转让给皮革制品制造商威廉·胡佛（William Hoover）。作为商人的胡佛看到了家用吸尘器市场的商机，于是以他的姓氏创建了吸尘器制造公司，将这一设计进一步改进后开始大规模生产家用吸尘器。“胡佛”逐步发展成为全球知名的吸尘器品牌，由于其影响力，英文 Hoover 成为“吸尘器”的别称。

从 19 世纪 60 年代末手动真空吸尘器出现，到电动真空吸尘器问世并迅速发展普及，真空吸尘器已经有 100 多年的发展历史。专利文献完整地记录了这一历程，清晰地展示了不同时期吸尘器的特征，特别是那些具有重要意义的突破与进步。而这正是专利文献作为人类发明创造宝库的珍贵价值——借助专利文献，人们得以跟上世界各地发明者创造的步伐，并站在前人的肩膀上开拓创新。

1901.GB 17433

——涉及地毯及其他材料灰尘提取的改进

N° 17,433  A.D. 1901

Date of Application, 30th Aug., 1901
Complete Specification Left, 28th May, 1902—Accepted, 17th July, 1902

PROVISIONAL SPECIFICATION.

"Improvements relating to the Extraction of Dust from Carpets and other Materials."

I, HUBERT CECIL BOOTH, of 5 Langham Chambers, Portland Place, in the County of London, Civil Engineer, do hereby declare the nature of this invention to be as follows:—

This invention has reference to the extraction of dust and similar impurities from carpets, rugs, curtains, bedding, upholstered furniture, the surfaces of walls and the like and has for its chief object to effect such extraction by passing over the surface of the carpet or the like, suitably formed hollow implements whose interior is maintained at a lower pressure than the atmosphere, by the action of a suction pump connected therewith by a pipe or conduit which also serves to conduct the extracted impurities along with the air to an appropriate contrivance or contrivances for separating and collecting them from the air so that the latter reaches the pump in a cleanly condition.

According to my invention the suction pump is actuated by an oil, steam, electric or other motor, the motor and the pump being both mounted on a portable frame or carriage which is adapted to stand outside the building wherein the dust extracting operation is to be performed. The said pump is connected by flexible piping with the dust extracting implement or implements which are formed with an opening or openings at the part contiguous to the surface of the carpet or the like, for the dust and other impurities to enter the interior of the implement and pass away through the said flexible piping by the suction produced by the pump. At suitable intervals in the length of the piping I provide the aforesaid contrivances for collecting the extracted impurities, such contrivances being preferably located at points outside the room wherein the dust extracting operation is taking place.

One of the collecting contrivances may comprise a casing containing water or other liquid through which the current of air charged with the extracted impurities passes from the lower part of said casing into a space at the upper part thereof above the level of the liquid and in so doing becomes filtered or deprived of its impurities and reaches the pump in a cleansed condition. In order to split up the air into fine streams or bubbles as it passes through the liquid and so assist in removing the impurities therefrom, I may provide perforated or reticulated partitions or screens near the bottom of the said casing in the path of the air; and for the purpose of abstracting the moisture from said air prior to its return to the pump, I may provide within the space within the casing above the level of the liquid, suitably arranged baffle plates or similar devices against which the moist air will flow in a circuitous course on its way to the pipe leading to the pump. The said casing may be furnished with a gauge glass for indicating the level of the liquid therein and may also be furnished with inlet and outlet openings controlled by valves or cocks for enabling the dirty liquid to be discharged therefrom and clean liquid to be supplied thereto. This contrivance may be situated on the aforesaid portable frame or carriage.

In addition to this contrivance another collecting contrivance may be employed which may comprise a box or casing having near its lower part a

[*Price 8d.*]

2 N° 17,433.—A.D. 1901.

Improvements relating to the Extraction of Dust from Carpets and other Materials.

series of inclined or other suitably arranged baffle plates and a dome shaped or other suitably shaped partition which divides the box or casing into two compartments. This partition may consist of some such material as cotton wool or linen and canvas and have rigidity imparted to it by a facing or facings of perforated or reticulated metal. The said box or casing may be made in two parts capable of being hermetically but detachably secured together and the edges of the partition may be held between frames or rings, one of which may project beyond the sides of the box or casing and lie between external flanges formed on the aforesaid two parts and connected together in any suitable manner. This construction permits of the said partition being readily disconnected from the box or casing when the parts of the latter are separated. The said baffle plates direct the heavier portions of the impurities towards the bottom of the lower compartment of the box or casing and prevent them from being unduly disturbed by air currents or when the said box or casing is opened for removing the impurities collected therein. Above the said partition I may provide a pipe having branch passages therein for the air to flow through and reach the pipe leading to the other collecting contrivance. The inlet pipe leading the air from the dust extracting implement to the lower compartment of said box or casing may be prolonged upwardly to a point a short distance below the said partition and may be furnished with a deflecting cap for preventing the air striking directly against said partition and for causing it to be more equally distributed within the space below said partition. It will not always be found necessary to employ both of the aforesaid dust collecting contrivances, in some cases only one of them need be used.

The dust extracting implements may be made in a variety of forms in accordance with the kind of materials they are intended to operate on. For carpets I may make the implement with a hollow base portion having an opening in its surface that lies next to the carpet and such opening may be capable of adjustment so as to regulate its area in accordance with requirements. A handle may be hinged to the said base for facilitating its movement over the surface of the carpet and the pipe for the air may pass along said handle and may be furnished with a cock or valve provided with means for readily enabling the latter to be actuated for opening the same. This valve is preferably so arranged that it normally assumes a closed position.

In some cases the hollow base portion may be formed or provided with a chamber to which the air pipe is connected, said chamber having a transverse or dome shaped partition composed of porous or other appropriate material serving as a separator of the dust or other impurities withdrawn from the carpet. The said partition may be provided with a spring controlled or other stem extending to the exterior of the chamber for the purpose of enabling said partition to be shaken or vibrated to remove therefrom the dust or other impurities that may collect thereon or therein. The inlet from the hollow base portion to the said chamber may be prolonged some distance into the latter and may be furnished with a flap or clack valve at its upper end.

When the dust extracting implements are of small size the aforesaid hinged handle may be dispensed with and a hollow handle be substituted therefor, the air which is sucked through the implement then passing through the handle which may be provided with a suitable cock or valve for opening the passage therein, such cock or valve normally remaining closed. An additional handle may if desired be provided at the front part of the implement for facilitating its movement.

If desired the face of the dust extracting implement may be mounted on adjustable wheels or rollers and springs may be provided between the said wheels or rollers and the implement. The surface of the implement above the surface of the carpet or other material to be cleansed may be slightly concave.

I am aware that it has before been proposed to employ dust extracting implements for removing dust and similar impurities from carpets and the like by

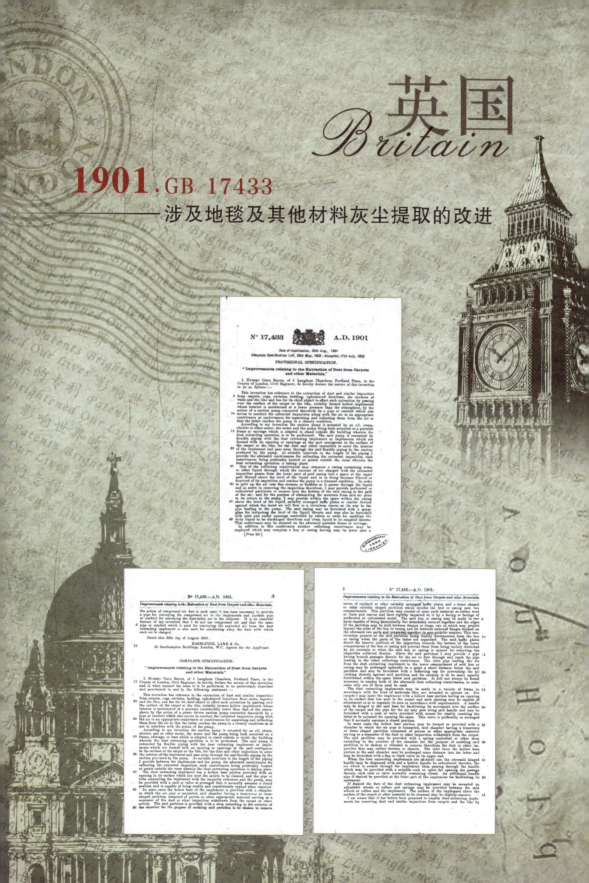

N° 17,433.—A.D. 1901. 3

Improvements relating to the Extraction of Dust from Carpets and other Materials.

the action of compressed air but in such cases it has been necessary to provide a pipe for conveying the compressed air to the implements and another pipe or conduit for conveying the dust-laden air to the collector. It is an essential feature of my invention that I do not use compressed air and that the same pipe or conduit which is used for conveying the aspirated air from the dust extracting implement is also used for conducting away the dust with which such air is charged.

Dated this 30th day of August 1901.

HASELTINE, LAKE & Co.
45 Southampton Buildings, London, W.C. Agents for the Applicant.

COMPLETE SPECIFICATION.

"Improvements relating to the Extraction of Dust from Carpets and other Materials."

I, HUBERT CECIL BOOTH, of 5 Langham Chambers, Portland Place, in the County of London, Civil Engineer, do hereby declare the nature of this invention and in what manner the same is to be performed, to be particularly described and ascertained in and by the following statement:—

This invention has reference to the extraction of dust and similar impurities from carpets, rugs, curtains, bedding, upholstered furniture, flour sacks, tapestry and the like, and has for its chief object to effect such extraction by passing over the surface of the carpet or the like, suitably formed hollow implements whose interior is maintained at a pressure considerably lower than that of the atmosphere, by the action of a power driven suction pump connected therewith by a pipe or conduit which also serves to conduct the extracted impurities along with the air to an appropriate contrivance or contrivances for separating and collecting them from the air so that the latter reaches the pump in a cleanly condition so as not to interfere with the action of the pump.

According to my invention the suction pump is actuated by an oil, steam, electric, gas or other motor, the motor and the pump being both mounted on a frame, carriage, or base which is adapted to stand outside or inside the building wherein the dust extracting operation is to be performed. The said pump is connected by flexible piping with the dust extracting implement or implements which are formed with an opening or openings at the part contiguous to the surface of the carpet or the like, for the dust and other impurities to enter the interior of the implement and pass away through the said flexible piping by the suction produced by the pump. At suitable intervals in the length of the piping I provide between the implements and the pump, the aforesaid contrivances for collecting the extracted impurities, such contrivances being preferably located at points outside the room wherein the dust extracting operation is taking place.

The dust extracting implement has a hollow base portion provided with an opening in its surface which lies next the article to be cleaned, and the pipe or tube connecting the implement with the impurity collectors and the pump, may be provided with a cock or valve so arranged that it normally assumes a closed position and is capable of being readily and expeditiously opened when required.

In some cases the hollow base of the implement is provided with a chamber to which the air pipe is connected, said chamber having a transverse or dome shaped partition composed of porous or other appropriate material serving as a separator of the dust or other impurities withdrawn from the carpet or other article. The said partition is provided with a stem extending to the exterior of the chamber for the purpose of enabling said partition to be shaken to remove

therefrom dust or other impurities. The inlet from the base portion to the said chamber may be prolonged and furnished with a flap or clack valve.

One of the impurity collectors may comprise a box or casing provided with a dome shaped porous partition. Beneath this partition is situated a deflecting cap upon which the impure air impinges, and a suitable outlet for the cleansed air is provided on the opposite side or above said partition. In the lower part of the collector there may be placed a series of inclined or other suitably arranged baffle plates for directing the heavier portions of the impurities toward the bottom of the box or casing.

Another impurity collector comprises a casing containing water or other liquid through which the current of impure air passes. Perforated or reticulated partitions or screens are arranged near the bottom of the casing in such manner as to split up the impure air into fine streams or bubbles as it passes through the liquid. Above the level of the liquid I provide suitable baffle plates to cause the air to flow in a circuitous course and be thereby dried before leaving the collector.

In order that my invention may be more clearly understood and readily carried into effect I will proceed to describe the same more fully in connection with the accompanying drawings in which:—

Figure 1 is a plan of the apparatus.

Figure 2 is a side elevation of an implement and its connections, partly in section.

Figure 3 is an under side view of the implement shown in Figure 2.

Figure 4 is a sectional side elevation of a modified form of implement.

Figure 5 is an under side view of the implement shown in Figure 4.

Figures 6 and 7 are a sectional side elevation and an under side view, respectively, of a further modification of the implement.

Figure 8 is a sectional elevation of the implement provided with a dust collecting chamber.

Figure 9 is an under side view of the implement shown in Figure 8.

Figures 10, 11 and 12 are vertical sectional views of the contrivances for collecting the impurities.

The same letters of reference indicate similar parts in the various figures.

In Figure 1, 1 is the hollow implement provided with a handle 2 and connected by flexible piping to a preliminary dust or impurity collector 3. The dust collector 3 is connected to a second dust collector 4 which communicates with the suction pump 5 actuated by the motor 6 driven by power from a generator or store 7; the dust collector 4, pump 5, motor 6 and generator 7 being carried by a suitable frame or base 8 mounted upon wheels.

In Figures 2 and 3, the implement 1 is provided with a base 9 having an orifice 10 which communicates with the hollow interior of the implement. The area of the orifice 10 can be adjusted by a plate 11 secured to the base 9 by screws 12 which pass through slots 13 in said plate. To adjust the area of the orifice 10, the screws 12 are loosened, the plate 11 then moved into the required position and the screws 12 tightened. The hollow interior of the implement 1 is connected to the dust collectors and the pump by flexible piping 14. 15 is a rotary valve which is rigidly connected to the handle 2 and is normally held closed by a spring rod 16 connected to the valve lever 17 by a pin and slot connection and retained in its closed position, as shown in the drawing, by a spring 18. To open the valve, the rod 16 is depressed against the resistance of the spring 18 by a hand lever 16ᵃ.

Referring to Figures 4 and 5 the hollow implement 1 is provided with a curved base 19 an orifice 20 and a handle 21. Between the flexible pipe 14 and the implement, is situated a tube 22 provided with a rotary valve 23. The lever 24 of the valve 23 is connected by a link 25 to one end of a spring 26 the other end of which is fixed to a lug carried by the tube 22. The spring tends to hold the valve 23 closed as shown in the drawing. When the implement is in

use, the operator grasps the tube 22 and the spring 26, drawing the spring 26 toward the tube 22 into the position shown in dotted line in Figure 4 and thereby opening the valve 23 and putting the implement in communication with the dust collectors and suction pump.

Figures 6 and 7 show an implement having a base 27, a handle 2 and a flexible pipe 14 similar to those shown in Figures 2 and 3. The base 27 is, however, provided with a non-adjustable orifice 28.

The implement shown in Figures 8 and 9 comprises a hollow dome-shaped chamber 29 having a similar base to that shown in Figures 2 and 3. The orifice 30 of this implement is prolonged and extends some distance within the chamber 29 and is provided with a flap-valve 31. Above the valve 31 is situated a transverse or dome-shaped partition or diaphragm of flexible resilient porous material 32 to filter the air drawn through the orifice 30 and valve 31 so that cleansed air leaves the chamber 29 by the pipe 14. A rod or stem 33 extends through the top of the chamber 29. One end of the stem 33 rests upon or is connected to the partition 32 and the other end of said rod is provided with a suitable push piece 34 situated outside the chamber 29. A spring 35 tends to hold the stem 33 in the position shown in the drawing. By pressing upon the push piece 34 and thereby depressing the stem 33, the partition 32 is also depressed. When the stem 33 is released, it is raised by the spring 35 and the partition 32 springs upwardly thereby shaking any dust or impurity collected on the under side of said partition, to the bottom of the chamber 29.

Referring to Figure 10, which is a vertical sectional view of the impurity collector 3, the said collecter comprises a box or casing having two parts 36 and 37, provided with flanges bolted together and having a packing ring 38 between them; the two parts 36, 37 being thus hermetically but detachably secured together. There is also situated between the parts 36, 37, a dome-shaped partition 39 of cotton-wool, linen, canvas or similar material made rigid by being faced with perforated or reticulated metal. The partition 39 is held in a groove 40 in the ring 38, so that the parts 36, 37 may be readily disconnected for removing the impurities from the part 37. The pipe 14 enters the part 37 and is prolonged upwardly to a point a short distance below the centre of the partition 39. 41 is a deflecting cap distributing the air drawn through the pipe 14, and preventing it striking directly against the partition 39. In the lower portion of the part 37, are arranged a series of inclined baffle plates 42 which direct the heavier particles of impurities, drawn in through the pipe 14, toward the bottom of the part 37 and prevent said particles from being unduly disturbed by air currents or when the casing is opened for removing the impurities collected therein. Air is drawn from the part 36 through a pipe 42 to the dust collector 4.

Figure 11 is a modification of the impurity collector shown in Figure 10. A dome-shaped porous partition of calico 53, is supported by a coiled spring 54 within the casing 55. The pipe 14 enters the casing 55 below the partition 53, and is provided with a deflecting cap 41 as above described. The casing 55 is provided, below the partition 53, with a hinged door 56 forming an outlet for impurities. The apparatus is provided with pressure gauges 57, 58. Strings 59 are connected to the top of the partition 53, whereby said partition may be drawn down against the resistance of the spring 54 when the door 56 is opened. Upon the release of said strings, the spring 54 will again raise the partition 53 thereby imparting to said partition a shock and shaking dust and impurities from it. Impure air enters the collecter through the pipe 14, the impurities are collected by the partition 53 and the cleansed air leaves the casing by the pipe 42. The inlet and outlet pipes in this form of impurity collector are carried by a single casting so as to avoid disconnecting the apparatus from the engine when it is required to open the collector for the removal of impurities or for repairs, or to replace the filtering medium.

The partition 39 in Figure 10 and also the partition 53 in Figure 11, is open

at the bottom, so that the impurities tend to fall by gravity away from the filtering medium.

Referring to Figure 12, which is a vertical sectional view of the dust collector 4, the pipe 42 enters the bottom of the collector which is partially filled with water or cleansing liquid. 43 is a perforated baffle plate situated immediately above the mouth of the pipe 42, and 44 is a reticulated partition extending across the collector above the baffle plate 43. The baffle plate 43 and partition 44 split up the air drawn into the collector through the pipe 42, into fine bubbles or streams as it passes through the liquid and so assist in removing impurities therefrom. 45 is a gauge for indicating the level of the liquid in the collector, 46 is the liquid inlet, and 47 is an outlet for liquid and impurities. Within the space above the liquid in the collector 4, are arranged baffle plates 48, 49 and 50 which cause the cleansed air to flow in a circuitous course and be thereby dried before entering a pipe 51 through perforations 52 and being thence conducted to the pump 5.

The dust extracting implements shown in Figures 2, 3, 6, 7, 8 and 9 are more particularly adapted for the treatment of carpets and the like, while the implement shown in Figures 4 and 5 is more particularly adapted for the treatment of upholstery.

The operation of the apparatus is as follows:—The under side of the implement 1 is passed over the surface to be cleansed, the valve 15 or 23 controlling the pipe 14 being opened and the suction pump 5 at work, a current of air and impurities extracted thereby from the article being cleaned is drawn through the orifice of the implement and thence through the pipe 14, and the impurity collectors 3, 4 to the pump 5 from which practically pure air is discharged.

If required the valve 15 or 23 may be omitted.

I find that it is essential to practical success to drive the pump, as above stated, by power, and to maintain a vacuum of at least 5 lbs. per square inch in the filter on the side of the filtering medium where the air and dust enter, when the apparatus is at work, and therefore it is only to extractors working with a considerable vacuum that my claims relate.

Having now particularly described and ascertained the nature of my said invention and in what manner the same is to be performed, I declare that what I claim in apparatus for extracting dust from carpets and other materials is:—

1. The combination of an extracting implement connected with a power driven suction pump, and dust collecting means interposed between the said implement and pump, substantially as and for the purpose specified.

2. In apparatus of the type claimed in the first claim a valve controlling the communication between the extracting implement and the impurity collectors and suction pump, said valve being so arranged that it normally assumes a closed position, substantially as described for the purpose specified.

3. In apparatus such as is claimed in the first claim a hollow extracting implement having a transverse vibratory porous diaphragm, an inlet orifice closed by a valve on one side of said diaphragm, and an outlet suction pipe on the other side of said diaphragm, substantially as described, for the purpose specified.

4. In apparatus such as is claimed in the first claim an impurity collector provided with a dome-shaped porous partition having beneath it a deflecting cap upon which the impure air is discharged, and having an outlet for cleansed air above said partition, substantially as described for the purpose specified.

5. In apparatus such as is claimed in the first claim an impurity collector containing liquid, in which are arranged perforated reticulated partitions for distributing the impure air, and having baffle plates arranged above the level of said liquid to cause the air to flow in a circuitous course before leaving the collector, substantially as described for the purpose specified.

6. An impurity extracting implement constructed arranged and adapted to operate substantially as described with reference to Figures 2 and 3, or to

Figures 4 and 5, or to Figures 6 and 7, or to Figures 8 and 9 of the accompanying drawings for the purpose specified.

7. An impurity collecting implement constructed, arranged and adapted to operate substantially as described with reference to Figure 10 or to Figure 11 or to Figure 12 of the accompanying drawings for the purpose specified.

8. Apparatus having its members constructed and arranged to co-operate substantially as described with reference to the accompanying drawings for the purpose specified.

Dated this 28th day of May 1902.

HASELTINE, LAKE & Co.
45 Southampton Buildings, London, W.C. Agents for the Applicant.

Redhill: Printed for His Majesty's Stationery Office, by Love & Malcomson, Ltd.—1902.

英国

Britain

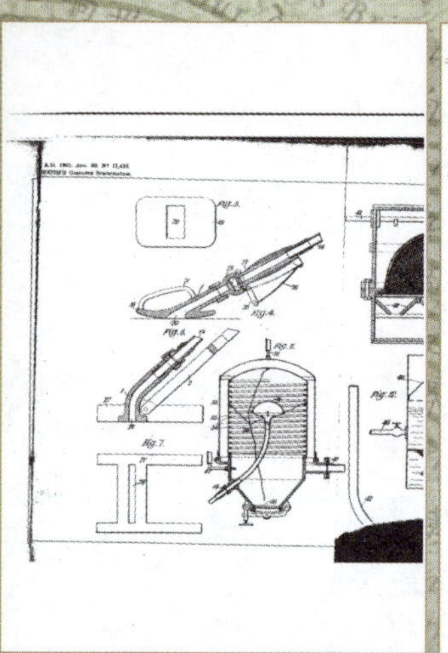

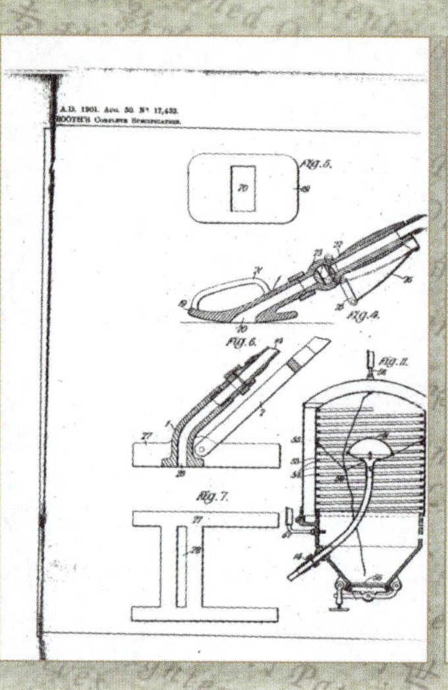

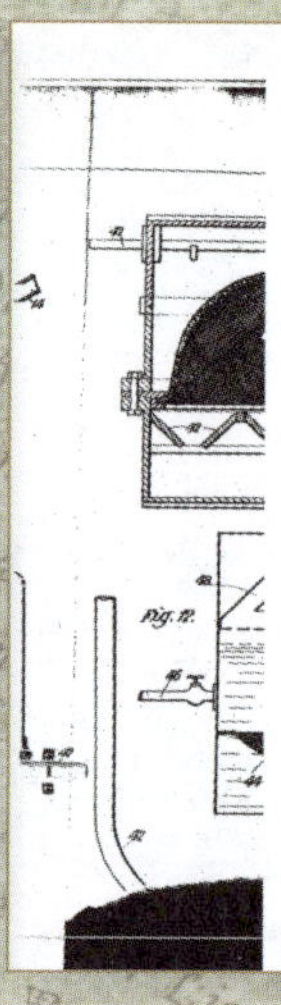

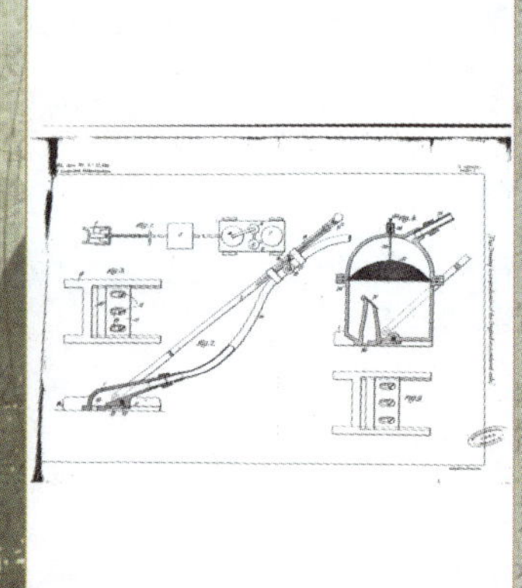

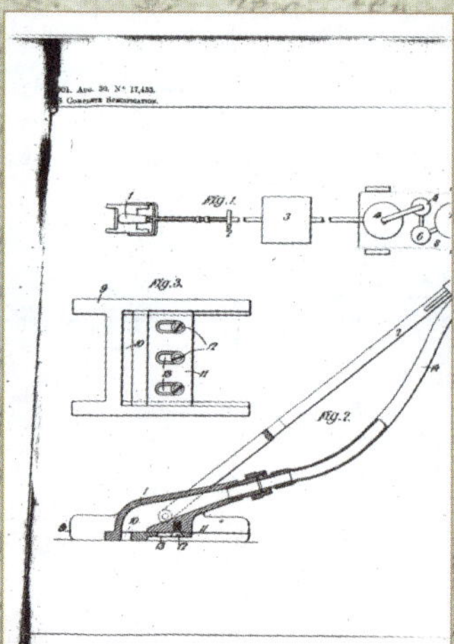

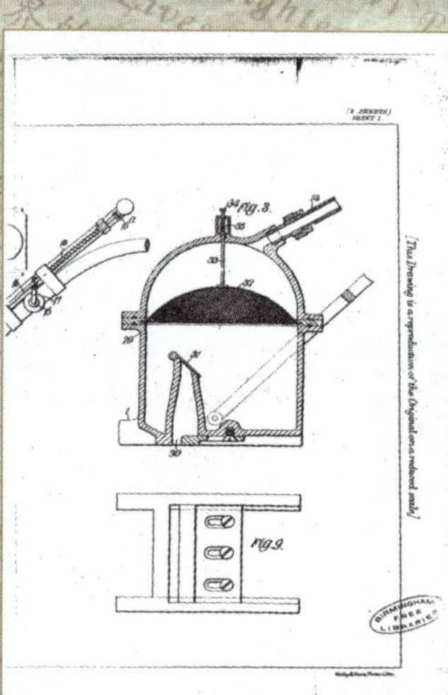

真空瓶 Dewar flask

詹姆士·杜瓦 James Dewar

詹姆士·杜瓦（1842–1923年），苏格兰化学家及物理学家。1893年，他发明了双层壁真空瓶，并获得了英国专利（GB439），该申请仅涵盖抽去空气形成真空的技术，而不是将其制成相应的容器。

1903年，当年被杜瓦雇用的两个德国吹玻璃工匠之一的赖因霍尔德·伯格，对杜瓦瓶进行了改进，在双壁形成的真空空腔中加入加固件，避免双层真空瓶壁因盛水等压力而爆破。该技术获得了名为“被两层真空壁包围的器皿”的德国专利（DE170057）。1906年，伯格又对该项技术进行了改进，并同时申请了德国、美国（US872795）和英国专利，名称为“具有壁间真空空腔的双层壁容器”。

杜瓦发明了储存低温液态气体的真空瓶（杜瓦瓶），伯格的专利发明对其进行了改进，形成了不同场合都能使用的保温瓶，例如，热水瓶、冰瓶等，并很快成为人类日常生活不可或缺的生活用品。

DE 170057

www.sipo.gov.cn

第九篇 真空瓶

真空瓶，又称为杜瓦瓶，以其发明人英国化学家和物理学家詹姆士·杜瓦爵士（James Dewar，1842-1923 年）的名字命名。早在古罗马时期人们就已经知道，双层容器能保暖，在庞贝城的废墟中就曾挖出过一个双层容器。真空瓶其实就是一个双层玻璃容器，两层玻璃胆壁都涂满水银，然后将两层壁间的空气抽掉，形成真空。两层胆壁上的水银可以防止辐射散热，真空能防止对流和传导散热，因此盛在瓶里的液体，温度不易发生变化，是储存液态气体、进行低温研究和晶体元件保护的一种较为理想的容器和工具。

詹姆士·杜瓦是剑桥大学教授和英国皇家学会会员，长期从事气体液化和低温现象研究。1880 年前后他开始研究液氧，1891 年他提出了液氧工业化生产的技术。为了保存这种液态氧，他发明了一种存储液态氧的容器，容器是双层玻璃的、胆壁上涂满水银，并且将两层壁间的空气抽掉形成真空。他将此项技术在英国申请了专利，并获得英国 GB439 号专利权，这一发明使他声名大噪，但是杜瓦的专利申请要求保护的范围仅涵盖抽去空气形成真空这一方法而不是将其制作成为相应的容器。他并没有意识到杜瓦瓶更广泛的用途，因此也没能从他的这项发明中获得利益。

杜瓦在试验中曾雇用了两名德国吹玻璃的工匠制作这种真空瓶。当年被杜瓦雇用的其中一位德国工匠赖因霍尔德·伯格（Reinhold

Burger）发现了商机，在杜瓦发明的基础上他对真空瓶进行了改造，制造出了我们日常生活中最常见的日用品——保温瓶。为了有效保护自己的发明，伯格于 1903 年申请了德国专利，名称为“被两层真空壁包围的器皿”，并于 1909 年获得德国 DE170057 号专利权，如图 9-1、图 9-2 所示。该专利较杜瓦的发明的改进之处在于，在双壁形成的真空空腔中加入加固件，避免双层真空瓶壁因盛水等压力而爆破。在这份专利说明书中还指出，该容器作为保温瓶等日常用具在使用时应配装类似军用水瓶这样的外壳。这一技术改进，将真空瓶带入了寻常百姓家庭。

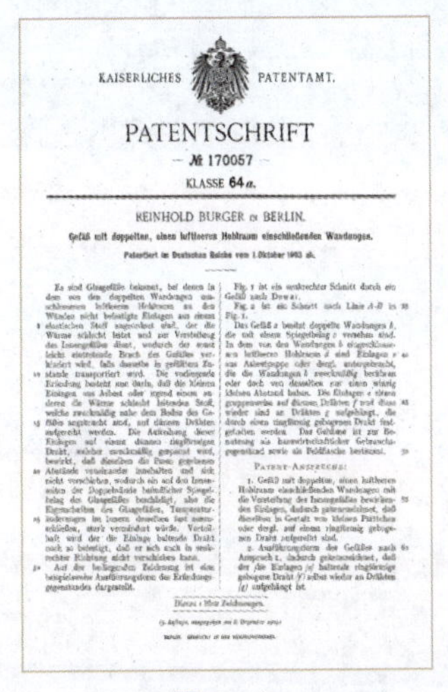

KAISERLICHES PATENTAMT.

PATENTSCHRIFT

— № 170057 —

KLASSE 64 *a*.

REINHOLD BURGER in BERLIN.

Gefäß mit doppelten, einen luftleeren Hohlraum einschließenden Wandungen.

图 9-1

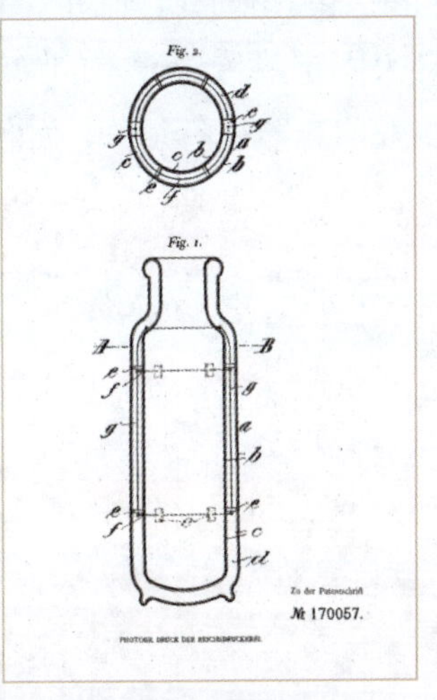

图 9-2

如这份专利说明书所展示的（图 9-2），下方为容器垂直剖视图，上方圆环为容器沿 A-B 截面的剖视图。容器的双层瓶壁具有镜面涂层，在双层瓶壁形成封闭的真空空腔的上下位置，设置有由石棉纸板或类似有弹性的导热性差的材料制成的衬垫，衬垫在空腔中与双层壁壁轻轻相接触，且左右各有两条细线穿成串并固定，一条细线是环状的，另一条细线是直线状的。

1904 年，伯格和另一德国吹玻璃工匠决定成立一家公司，专门生产这种真空瓶。为此，他们还专门就这种真空瓶的品名举行了一场公开竞标活动。在活动中，一个慕尼黑居民提出从希腊词“therme”

（热）衍变成“thermos”来命名这种真空瓶，即热水瓶。其后伯格将“Thermos”注册登记为商标。

由于伯格发明的这种真空瓶，仅相当于我们现在常见的热水瓶胆，其市场推广性还不是很理想。因此公司成立之后，为了生产出更受市场欢迎的产品，伯格不断改进技术。1906 年，他以“具有壁间真空空腔的双层壁容器”这一名称同时向德国、美国、法国和英国申请了专利。为了便于解读专利文献我们通过同族专利检索查到这项发明于 1907 年 12 月获得美国 US872795 号专利权。从这份美国专利说明书中我们了解到，伯格这项技术的发明点是设置于双层壁间的加固垫块不再用细线和细线环固定，代之以玻璃瓶壁的不同凹、凸方式或凹、凸环来固定热传导性差的弹性材料，如石棉、软木垫块或环，从而对真空双层容器进行加固，如图 9–3、图 9–4、图 9–5 所示。

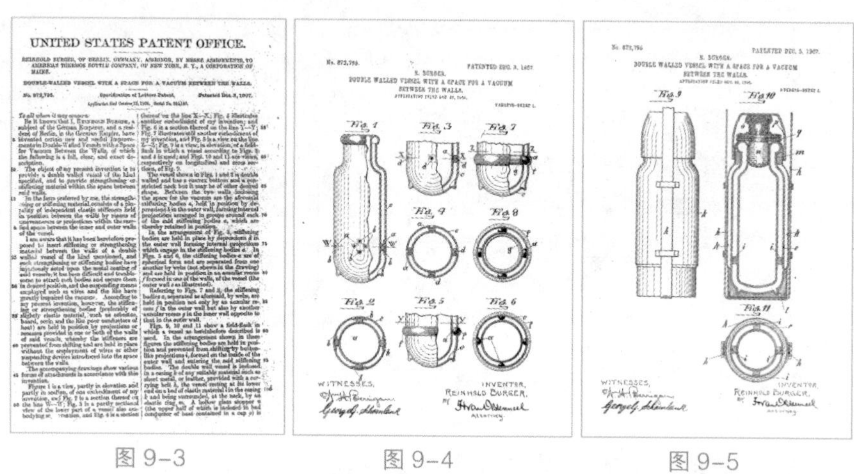
UNITED STATES PATENT OFFICE.

DOUBLE-WALLED VESSEL WITH A SPACE FOR A VACUUM BETWEEN THE WALLS.

Specification of Letters Patent. Patented Dec. 3, 1907.

图 9–3　图 9–4　图 9–5

如这项专利说明书附图所示（图 9–5），一种野外瓶的侧视图，分别为该瓶的垂直剖视图，以及为在设有垫块处截面的剖视图。加固块由凸出部分定位，防止该加固块产生位移，从而形成内外壁之间的加固作用。凸出部分是在玻璃双层壁容器吹制中形成的。真空双层壁容器具有诸如用金属板或皮革制成的外壳，该外壳装有提带，真空双层壁容器的底部坐落在外壳底面上设置的弹性衬座上，真空双层壁容器的颈部与外壳之间围填有弹性环。真空双层壁容器口设有中空玻璃塞头，塞头上半部的帽中为热传导性差的材料，塞头可借助于盖往下

压。这个盖设有内螺纹与外壳上部的外螺纹吻合连接，通过螺旋调节盖的下旋位置，从而调节塞头的下压力，弹性衬垫使塞头压紧在真空双层壁容器口的上缘上，使之与外界隔绝。盖还可作为杯子使用。

这些对真空瓶的技术改进，使伯格的 Thermos 公司的产品有了更为广大的市场。Thermos 公司在随后的几十年间申请了 100 多件与真空瓶相关的专利，专利申请国家主要集中在工厂所在地——美国、英国、加拿大，在德国和瑞士有少量申请。20 世纪 20 年代末，Thermos 公司实现了真空瓶的工业化生产，相继获得了多项专利，例如名称为“双层玻璃容器的加工设备”的专利发明，1929 年获得美国 US1735027 号专利权、1930 年获得法国 FR673216 号专利权。Thermos 公司还致力于真空瓶的多种应用，例如用于盛装冰块的大口型保温容器，如图 9-6、图 9-7 所示，于 1937 年获得名为“隔热容器”的美国 US2102164 号专利权。

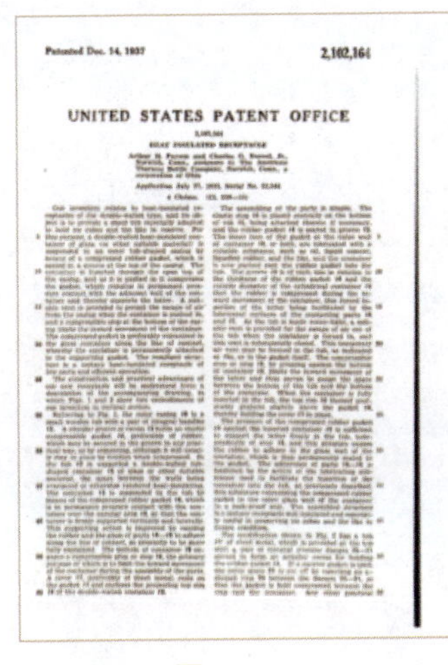

Patented Dec. 14, 1937　　2,102,164

UNITED STATES PATENT OFFICE

图 9-6

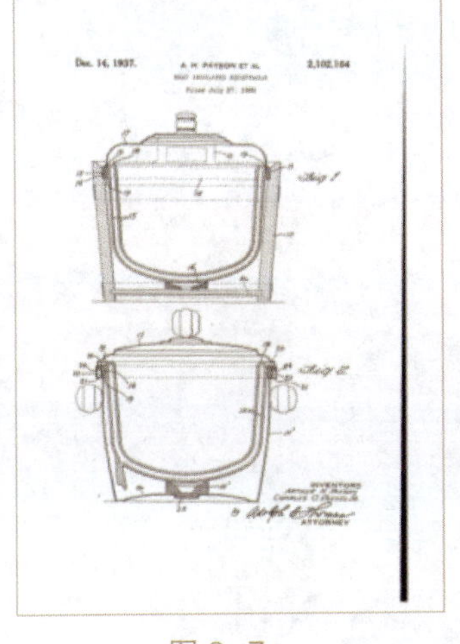

图 9-7

一百多年过去了，詹姆士·杜瓦的发明仍在实验室中被广泛使用，实验人员用杜瓦瓶来装运、储存低温液体。而伯格对杜瓦发明的改进使真空瓶的功能更为多样，得到更为广泛的应用，成为人类日常生活中不可或缺的生活用品。

1903.DE 170057

——被两层真空壁包围的器皿

KAISERLICHES PATENTAMT.

PATENTSCHRIFT

— № 170057 —

KLASSE 64a.

REINHOLD BURGER in BERLIN.

Gefäß mit doppelten, einen luftleeren Hohlraum einschließenden Wandungen.

Patentiert im Deutschen Reiche vom 1. Oktober 1902 ab.

— № 170057 —

KLASSE 64a.

REINHOLD BURGER in BERLIN.

Gefäß mit doppelten, einen luftleeren Hohlraum einschließenden Wandungen.

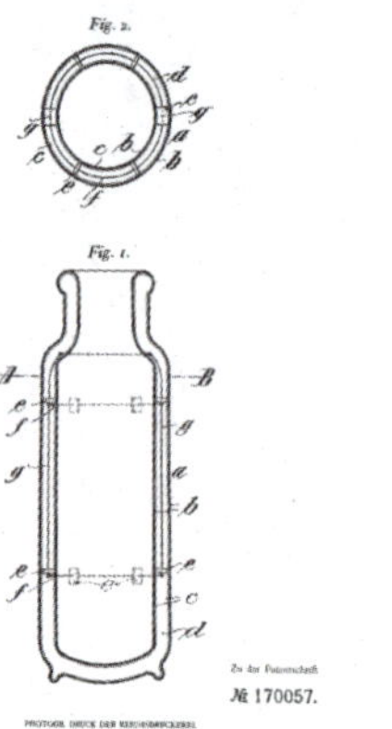

Zipper invention 拉链的发明

A.D. 1915

吉德昂·逊德巴克
Gideon Sundback

吉德昂·逊德巴克（Gideon Sundback，1980–1954年），美籍瑞典发明家。1880年出生于瑞典，是一名电气工程师，从小就对机械感兴趣，1905年移民到美国，不久开始从事拉链的发明创造。1906～1914年，在为Talon有限公司工作时，逊德巴克在拉链开发方面取得了重大进展，并申请了专利。

1914年8月27日申请、1915年首先公布的英国GB12261号专利是现代拉链的起源，为拉链问世和发展奠定了基础。根据它开合时发出的摩擦声，人们给它取了一个形象的名字，叫做“ZIPPER”，也就是“拉链”。

最初拉链主要用于靴子和烟草袋，20年后开始在时尚产业界流行。

GB 12261

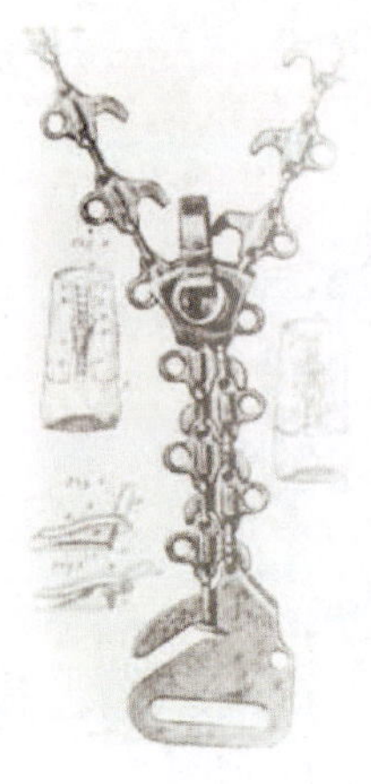

第十篇 拉 链

拉链可以说是现代人最为熟悉的物件之一。而它在人们生活中出现可以追溯至 100 年前。那时，人们的衣服和靴子基本采用扣子或绑带来固定，穿脱起来非常麻烦。于是，有人开始思考，若是能有更简便的办法把所有扣子都连起来，能够更快地扣上和解开，那肯定会大大节省时间和精力。人们开始尝试用一连串钩环式的扣件取代纽扣和绑带，对于拉链的研制和应用从那时拉开了序幕。

1893 年，美国工程师惠特科姆 · L. 贾德森（Whitcomb L. Judson）发明了一种“用于鞋子的钩环式锁扣”，并获得了美国 US504038 号专利权，如图 10-1 所示，这便是拉链的雏形。在这项发明中，接缝一侧有一串钩子，另一侧有一串钩眼，通过拉动拉头沿接缝上下活动

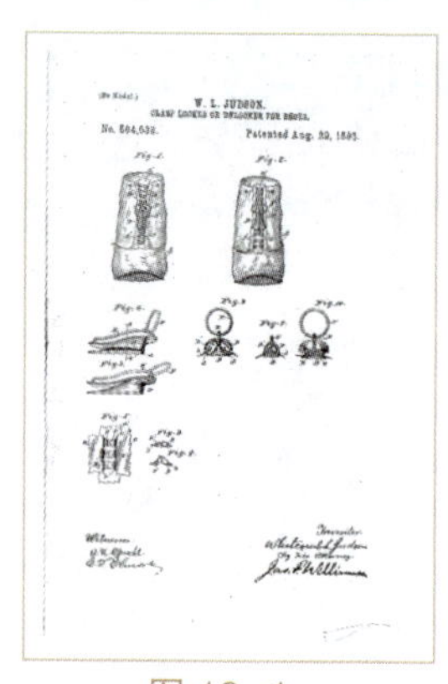

图 10-1

使两边对应的钩子和钩眼互相钩住或解开，从而实现接缝处的扣合和解扣。

贾德森在芝加哥世界博览会上向公众展示了带有这种锁扣的鞋，引起了人们的关注。贾德森便成立了公司来生产这种锁扣。但是这种锁扣使用起来有些笨重，扣在一起的钩环很容易松开，拉头移动也不太顺畅，制造工艺又比较烦琐。这一系列原因使得这一发明没能流行起来。

贾德森的公司继续对这种原始的“拉链”进行改进。1913 年，该公司的一位瑞典籍工程师吉德昂·逊德巴克（Gideon Sundback，1880–1954 年）对贾德森的发明进行了重要改进，成功克服了上面所提到的一系列缺陷，终于开发出一种使用起来流利和顺手的拉链。1914 年，逊德巴克为他的发明在美国提交了名为“用于服装及其他用途可分离紧固件的改进”的专利申请，又以此作为优先申请，于 1915 年分别向英国和德国提出了专利申请。由于三个国家专利审批制度和进度的差异，这件申请首先于 1915 年在英国公布（GB12261 号），随后又相继于 1917 年、1920 年获得了美国 US1219881 号和德国 DE325390 号专利权。通过最先公布的这篇英国专利文献，如图 10–2、图 10–3 所示，我们可以看到，逊德巴克将贾德森发明中的钩环改进为可以交错咬合的链牙（也称为咬

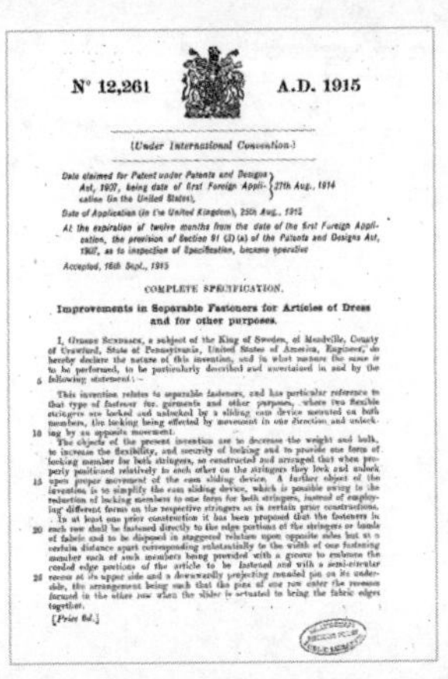

N° 12,261　　A.D. 1915

(Under International Convention.)

Date claimed for Patent under Patents and Designs Act, 1907, being date of first Foreign Application (in the United States), 27th Aug., 1914

Date of Application (in the United Kingdom), 25th Aug., 1915

At the expiration of twelve months from the date of the first Foreign Application, the provision of Section 91 (3) (a) of the Patents and Designs Act, 1907, as to inspection of Specification, became operative

Accepted, 16th Sept., 1915

COMPLETE SPECIFICATION.

Improvements in Separable Fasteners for Articles of Dress and for other purposes.

I, Gideon Sundback, a subject of the King of Sweden, of Meadville, County of Crawford, State of Pennsylvania, United States of America, Engineer, do hereby declare the nature of this invention, and in what manner the same is to be performed, to be particularly described and ascertained in and by the following statement:—

This invention relates to separable fasteners, and has particular reference to that type of fastener for garments and other purposes, where two flexible stringers are locked and unlocked by a sliding cam device mounted on both members, the locking being effected by movement in one direction and unlocking by an opposite movement.

The objects of the present invention are to decrease the weight and bulk, to increase the flexibility, and security of locking and to provide one form of locking member for both stringers, so constructed and arranged that when properly positioned relatively to each other on the stringers they lock and unlock upon proper movement of the cam sliding device. A further object of the invention is to simplify the cam sliding device, which is possible owing to the reduction of locking members to one form for both stringers, instead of employing different forms on the respective stringers as in certain prior constructions.

In at least one prior construction it has been proposed that the fasteners in each row shall be fastened directly to the edge portions of the stringers or bands of fabric and to be disposed in staggered relation upon opposite sides but at a certain distance apart corresponding substantially to the width of one fastening member each of such members being provided with a groove to embrace the corded edge portions of the article to be fastened and with a semi-circular recess at its upper side and a downwardly projecting rounded pin on its underside, the arrangement being such that the pins of one row enter the recesses formed in the other row when the slider is actuated to bring the fabric edges together.

[Price 8d.]

图 10–2

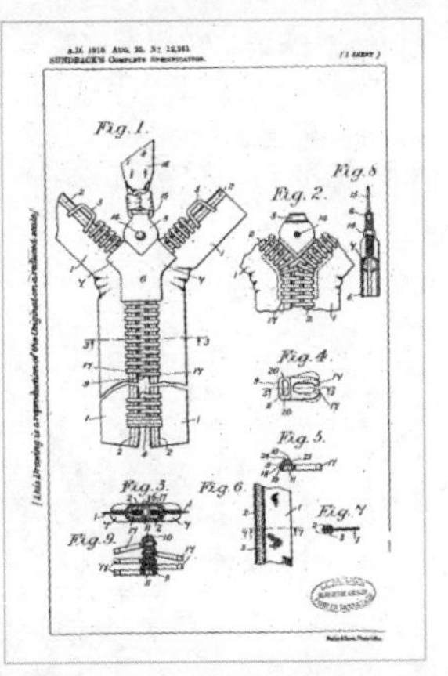

图 10–3

齿），又在最前端和最末端的链牙处添加了限位挡头。逊德巴克的改进使得拉链更加牢固，只有拉头滑动才能松开咬合的链牙，前后挡头有效地限制和固定了拉头——这正是今天我们熟悉的拉链的基本设计。

逊德巴克将他发明的拉链命名为“泰龙”（Talon）。“泰龙”一上市就受到了欢迎，贾德森的公司收到了几千份订单。1913 年“泰龙”被用于美国海军的制服，后来又被用到手套和烟草袋上。1937 年，贾德森的公司也更名为“泰龙”。今天，“泰龙”已经是世界最为知名的拉链品牌之一，并以制造出世界上第一条拉链和世界首家拉链公司而著称。

英文中“拉链”被称为“zipper”，这个词是从何而来呢？1923 年，美国古德厘奇公司在其生产的橡胶雨靴上使用了逊德巴克发明的拉链。由于它在开合时会发出“滋滋”的摩擦声，古德厘奇公司就给它取了一个形象的名字“ZIPPER”，这个朗朗上口的词后来就被人们用来指代拉链了。

贾德森和逊德巴克的发明奠定了拉链制造的基础。通过检索专利文献我们发现，直到 21 世纪初，依然有研究者在前人的构思基础上研制新的拉链。例如下列 3 件专利，如图 10-4、图 10-5、图 10-6 所示，这是一家美国公司在 2002 ~ 2004 年的 3 项关于拉链的发明，从这份

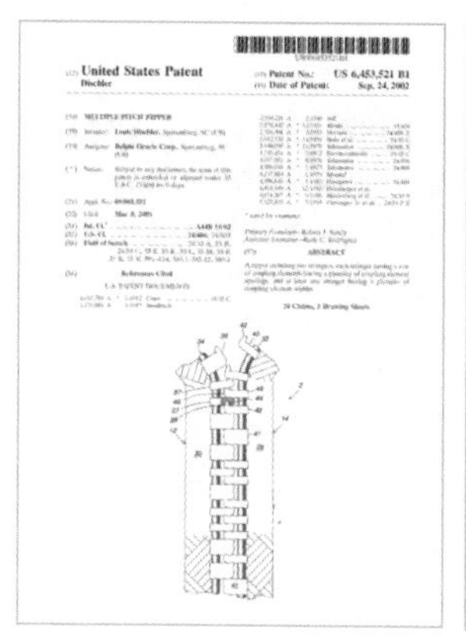

United States Patent
Dischler
Patent No.: US 6,453,521 B1
Date of Patent: Sep. 24, 2002

图 10-4

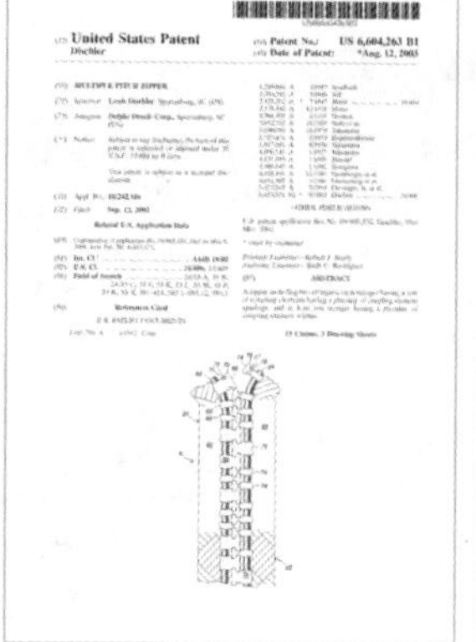

United States Patent
Dischler
Patent No.: US 6,604,263 B1
Date of Patent: *Aug. 12, 2003

图 10-5

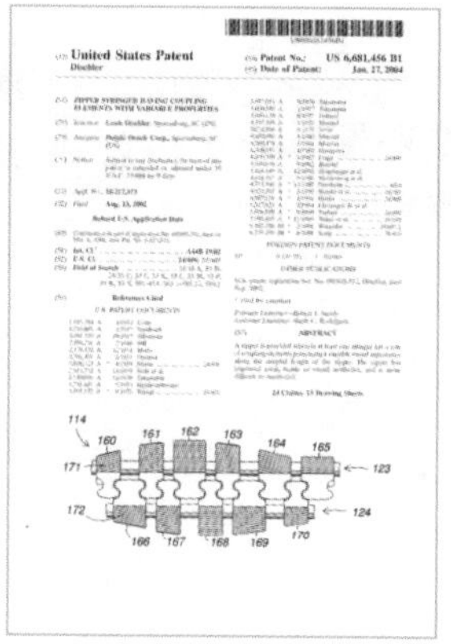

United States Patent
Dischler
Patent No.: US 6,681,456 B1
Date of Patent: Jan. 27, 2004

图 10-6

专利说明书可看出，该公司的发明沿用了逊德巴克的拉链的基本结构，又对拉链咬齿间距、所用材料等方面进行了改进。

由于巨大的实用性，拉链在用于靴子、手套和烟草袋等日用品后，又在时尚产业界流行。今天，拉链的使用已经不仅仅局限于服装鞋帽业，在人们生活中处处可以看到拉链的身影，它甚至被应用到外科手术缝合和食品保鲜包装上。

从百年前麻烦地解扣子、系绑带，到今天轻松地拉动拉链，人类文明的进步就是由无数个这样的发明创造所推动的。贾德森和逊德巴克也许面对细小的钩环、链牙试验了千百次，才成功发明了拉链。同时，他们重视通过专利对自己的发明进行保护。专利制度为发明家在商业上的成功提供了保障，为他们坚持不懈地投身发明创造和不断改进突破起到了激励作用。

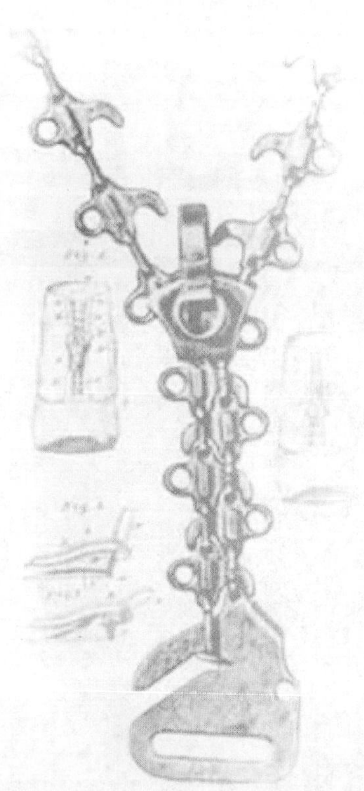

1915. GB 12261

用于服装及其他用途可分离紧固件的改进

N° 12,261 A.D. 1915

(Under International Convention.)

Date claimed for Patent under Patents and Designs Act, 1907, being date of first Foreign Application (in the United States), 27th Aug., 1914

Date of Application (in the United Kingdom), 25th Aug., 1915

At the expiration of twelve months from the date of the first Foreign Application, the provision of Section 91 (3) (a) of the Patents and Designs Act, 1907, as to inspection of Specification, became operative

Accepted, 16th Sept., 1915

COMPLETE SPECIFICATION.

Improvements in Separable Fasteners for Articles of Dress and for other purposes.

I, Gideon Sundback, a subject of the King of Sweden, of Meadville, County of Crawford, State of Pennsylvania, United States of America, Engineer, do hereby declare the nature of this invention, and in what manner the same is to be performed, to be particularly described and ascertained in and by the following statement:—

This invention relates to separable fasteners, and has particular reference to that type of fastener for garments and other purposes, where two flexible stringers are locked and unlocked by a sliding cam device mounted on both members, the locking being effected by movement in one direction and unlocking by an opposite movement.

The objects of the present invention are to decrease the weight and bulk, to increase the flexibility, and security of locking and to provide one form of locking member for both stringers, so constructed and arranged that when properly positioned relatively to each other on the stringers they lock and unlock upon proper movement of the cam sliding device. A further object of the invention is to simplify the cam sliding device, which is possible owing to the reduction of locking members to one form for both stringers, instead of employing different forms on the respective stringers as in certain prior constructions.

In at least one prior construction it has been proposed that the fasteners in each row shall be fastened directly to the edge portions of the stringers or bands of fabric and to be disposed in staggered relation upon opposite sides but at a certain distance apart corresponding substantially to the width of one fastening member each of such members being provided with a groove to embrace the corded edge portions of the article to be fastened and with a semi-circular recess at its upper side and a downwardly projecting rounded pin on its underside, the arrangement being such that the pins of one row enter the recesses formed in the other row when the slider is actuated to bring the fabric edges together.

[Price 6d.]

2 N° 12,261.—A.D. 1915.

Improvements in Separable Fasteners for Articles of Dress and for other purposes.

According to the present invention, the stringers are alike, as in some prior types of fasteners and preferably consisting of the known fabric tape provided with a beaded or corded edge, upon which the locking members are clamped. The locking members are all alike, and therefore interchangeable, and in general form consist of contractible jaw portions which are clamped upon the tape and projecting locking portions of elongated cup shape, so that the outside of one member nests within the recess of an adjoining member when in locked relation. Consequently, it will be seen that the members on one stringer alternate with those on the other, so that when the sliding operating device is moved back and forth, the locking members will be engaged and disengaged according to the direction of movement. A further and one of the essential features of the invention resides in the shape and configuration of the locking members, these being as shown in enlarged scale on the drawings provided with exterior and interior rounded surfaces, and are somewhat elongated transversely.

This construction will provide an adequate opening for the locking movement and yet prevent the liability of catching while guiding the projecting part smoothly and easily into the locking recess. Thus a snug fit is obtained and at the same time ample provision is given for movement of one on the other without coming out when the fastener is flexed transversely. At the same time this construction gives facility for relative longitudinal movement without disengagement.

In the accompanying drawings:—

Figure 1 is a perspective view on an enlarged scale of a fastener embodying the invention;

Figure 2 is a detail view showing the manner of locking and unlocking;

Figure 3 is a cross section on the line 3—3 of Figure 1;

Figure 4 is a detail view of one of the locking members showing the recess;

Figure 5 is a cross section thereof on the line 5—5 of Figure 4;

Figure 6 is a detail view of one of the stringers;

Figure 7 is a cross section thereof on the line 7—7 of Figure 6;

Figure 8 is a cross section of the slider; and

Figure 9 is an enlarged detail of the locking members.

1—1 represent stringers which preferably will be made of woven tape having sewed thereto on opposite sides cords 2—2 to form a beaded edge. 3 represents the line of stitching as employed in the preferred construction. Each of these stringers may be secured together at one end by fastening links 4, and will have at the opposite end a stop member 5. 6 represents the sliding cam operating device, which consists of two stamped sides 7 each having a Y-shaped channel, with their sides separated to permit passage of the tape portion of the stringers 1. 8 represents a cap having a point to fit between the diverging channels, and secured to the slider by a rivet 14. 15 is a ring in which may be placed an operating tape or bow 16. The stampings 7 may be integral and doubled over or they may be separate, in either event a rigid structure being provided when the cap 8 is applied thereto and the rivet 14 set.

Figure 4 represents one of the locking members, which consists of jaws 17—17 having a base portion 9 which is stamped or otherwise formed to provide an external locking projection 10 and an internal recess 11. These members, as well as the locking members 4 and the stop members 5 are stamped out of flat metal, and the jaws 17, 17 of the locking members are spread apart as shown in dotted lines in Figure 4 so as to pass over the corded edges 2 of the stringers. Then the jaws are set by a die press or other suitable means, the jaws thereby becoming firmly clamped to the stringers by the compression of the corded edges 2. This will be seen in Figure 3, where the corded edges have been distorted by the setting of the locking members thereon so as to completely fill the slot in the locking members, and the jaws 17, 17 engaging firmly on the tape immediately behind the corded edges.

This distortion of the corded edge also produces slight elongation and this

N° 12,261.—A.D. 1915. 3

Improvements in Separable Fasteners for Articles of Dress and for other purposes.

can be compensated for by feeding the cord slower than the tape when sewing same together.

From the assembly in Figure 1, it will be seen that the members 17 on one stringer alternate with those on the other stringer, and consequently it is possible to make the members on both stringers alike and cause them to lock and unlock readily, and remain locked even while the fastener is almost doubled on itself. At the same time, owing to the rounded and transversely elongated shape of the projections and recesses, the fastener is very flexible without being loose. Flexibility is also increased by reason of the relatively large number of locking members provided, which is possible because these members are thin and their projections and recesses can be proportioned so that one will not touch another when the fastener is bent transversely. This is an important consideration in fasteners of this type. Thus it will be seen that the shape of the projections and recesses is such that when engaged, the stringers have practically no movement of separation, but yet the engagement is secure without being stiff, because the locking members on one stringer can rock or oscillate freely relatively to those on the other stringer without disengagement.

Practical success in a fastener of this character is largely dependent upon proper construction of the locking members. Not only must these members remain engaged when the fastener is flexed transversely or one stringer moved longitudinally relatively to the other, but the locking members must guide smoothly into and out of each other without liability of jamming on the slider or on each other, and must always lock without objectionable looseness. These objects are all obtained in the fastener of this invention, wherein not only are the locking members for both stringers alike, but they lock easily and surely without jamming and yet provide the necessary flexibility. The members 10 are provided with opposite inclined guiding surfaces 18, 21, and 20, 20, the surfaces 21 being somewhat steeper, and rounded at the corners so as to fit in the sockets 11 shown in Figure 4. The upper surfaces of the locking projections are slightly rounded as shown at 24, and the projections and sockets are also elongated transversely as shown in Figure 4, and the under surface may be set at an angle to the surface of the jaws 17, so as to form a cam surface between the front edge of the socket 11 and the sharp edge 19, over which the rounded heads 24 may slide in locking and unlocking. The angle of the guiding surface 18 conforms to the curve of the slider, and this surface, in conjunction with the edge 19, causes the adjacent members on the opposite stringers to ride one on the other, and guide each other, into proper locking engagement. Thereby catching is eliminated, and by reason of the members guiding each other into locking engagement, the radius of the slider channel can be increased. This is of advantage when the fastener is used on stiff garments such as corsets. The projections and sockets fit closely, and thus render the distance between the jaws independent of their thickness, which enables proper locking to be secured with jaw members thinner than the projections, and consequently more separated on the stringers. Greater spacing of the jaw members increases the flexibility, but without the elongated and conical shaped projections and recesses herein shown, this increased flexibility would enable the members to be easily disengaged by transverse bending.

From the foregoing it will be seen that by the present invention an improved fastener has been produced which is much more simple and cheap to manufacture than others heretofore invented, not only are both stringers as well as the respective members thereon alike, but all the members themselves can all be made by simple die operations from flat material. The only assembling operations are those required to attach the members to the tapes. The finishing of the locking members such as smoothing or tumbling is very simple, because all are alike and the dies can be so designed as not to require much subsequent finishing of these members. It will be understood that the connecting members 4 and the stop members 5 may have jaws 17 on each end similar to

4 N° 12,261.—A.D. 1915.

Improvements in Separable Fasteners for Articles of Dress and for other purposes.

those of the locking members, which are only on one end. As shown herein, it will be seen that the locking members are of a rounded truncated conical shape, the tops being cut off and rounded so as not to leave projecting edges which would interfere with their engagement and disengagement.

Having now particularly described and ascertained the nature of my said invention, and in what manner the same is to be performed, I declare that what I claim is:—

1. A separable fastener of the class having two flexible tapes each carrying an interlocking socket and head member characterised by the fact that the sockets on one side of each member and the heads on the opposite side are transversely elongated and the exterior surfaces of the heads constitute inclined guiding surfaces for the co-operating socket member with which it is to engage.

2. In separable fasteners as set out in Claim 1 projecting claw ends integral with the socket and head member and adapted when clamped to the rounded edge of the fabric to distort and flatten same.

3. In separable fasteners as set out in the preceding claims a sliding cam operating device engaging with the interlocking socket and head members and stop members at one or both ends of the stringers.

4. Separable fasteners constructed and adapted to operate substantially as described and as illustrated in the accompanying drawings.

Dated this 25th day of August, 1915.

PHILIP M. JUSTICE,
Chartered Patent Agent,
London,
For the Applicant.

Printed for His Majesty's Stationery Office, by Love & Malcomson, Ltd.—1915.

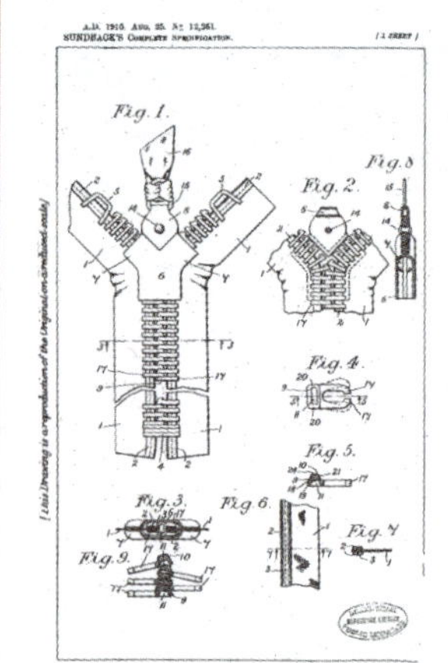

交通信号灯的发明
Traffic Signal

加勒特·摩根
Garrett Morgan

第一次对交叉路口交通的控制尝试起源于1868年英国伦敦，当时由警察手工轮流变换指挥用的旗帜，称之为旗语。1914年在美国俄亥俄州的克里夫兰市出现了第一台电力驱动的交通信号灯。1923年加勒特·摩根发明了一种可以通过机械连接远距离手动操纵“停、走、全停”定位功能的T形柱交通信号灯，并获得美国专利（US1475024），成为后来各种类型交通信号灯的基础。

摩根（1877-1963年），著名的非裔美国发明人。他的发明除交通信号灯外，还有呼吸保护面罩和毛发拉直方法。2002年被Molef Kete Asante列为最伟大的100位非裔美国人。

GB 1475024

G. A. MORGAN
TRAFFIC SIGNAL
Filed Feb. 27, 1922
1,475,024
2 Sheets-Sheet 1

第十一篇 交通信号灯

交通信号灯是我们每天经过十字路口的时候都会看到的，“红灯停、绿灯行”是大家从小就熟稔于心的交通规则。那么交通信号灯到底是谁发明的，是从什么时候开始出现的呢？

19 世纪初，在英国中部的约克城，女性通过不同颜色的着装来表示不同的身份。其中，着红装的女人表示已婚，而着绿装的女人则是未婚。后来，人们受到红绿装的启发，将其应用于交通控制中。1868 年英国伦敦第一次尝试对交叉路口的交通进行管控，当时由警察手工轮流变换指挥旗帜，称为旗语。

随着第一批美国制造汽车的出现，20 世纪初的美国城市，经常会见到自行车、畜力车、汽车和行人共用一条街道和公路的情形，交通事故频繁发生。许多早期的交通信号灯就在这一时期相继出现了。1910 年，芝加哥的厄内斯特·西林（Earnest Sirrine）申请的美国 US976939 号专利也许是第一个自动交通管控系统，该系统使用无照明的“停止”和“继续”作为指挥信号。

1914 年在美国俄亥俄州的克里夫兰市出现了第一台电力驱动的交通信号灯——“电气信号灯”。随着各种交通工具的发展和交通指挥的需要，第一盏名副其实的三色灯（红、黄、绿三种标志）于 1918 年诞生，它是三色圆形的四面投影器，被安装在纽约市第五大道的一座高塔上，由于它的诞生，城市交通大为改善。

值得一提的是黄色信号灯的发明者是我国的胡汝鼎，当年他怀着科学救国的抱负到美国深造，在美国通用电器公司任职。一天，他站在十字路口等待绿灯信号，当他看到绿灯正要通过时，一辆转弯的汽车擦身疾驰而过，险些酿成事故。回到宿舍，他反复琢磨，终于想到在红灯与绿灯中间再加上一个黄色信号灯，提醒人们注意危险。他的建议立即得到有关方面的肯定。于是红、黄、绿三色信号灯成为一个完整的指挥信号家族，遍及全世界的交通领域了。中国最早的马路红绿灯，于 1928 年出现在上海的英租界。

1923年，加勒特·摩根（Garrett Morgan，1877–1963年）发明了一种成本低廉的可以通过机械连接远距离手动操纵“停、走、全停”定位功能的T形柱交通信号灯。摩根于1922年以“交通信号灯”为名提出专利申请，1923年获得美国US1475024号专利权，如图11–1、图11–2、图11–3所示。在他的发明中对“交通信号灯”描述为：“本发明涉及交通信号，特别是适合于两个或两个以上的街道相邻交叉路口，并且能手动操作用于指挥交通……另外，我发明的这一个交通信号灯，可以很容易和便宜地制造出来。”与早期的交通信号灯相比，摩根的交通信号灯更适合批量生产和应用。

摩根的交通信号灯是一个 T 形结构的立柱，能够发出三种交通指

Patented Nov. 20, 1923. 1,475,024

UNITED STATES PATENT OFFICE.

图 11–1

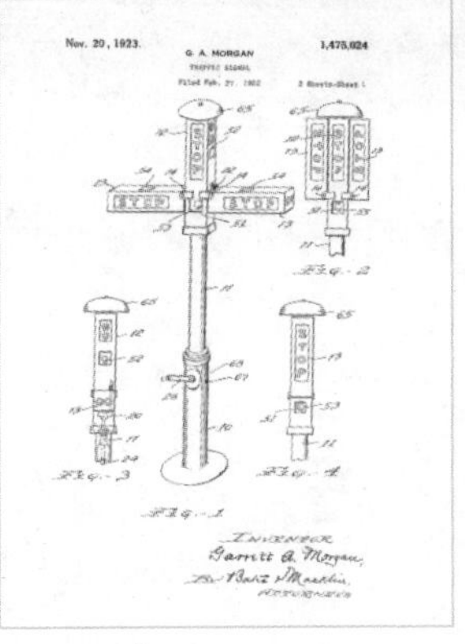

图 11–2

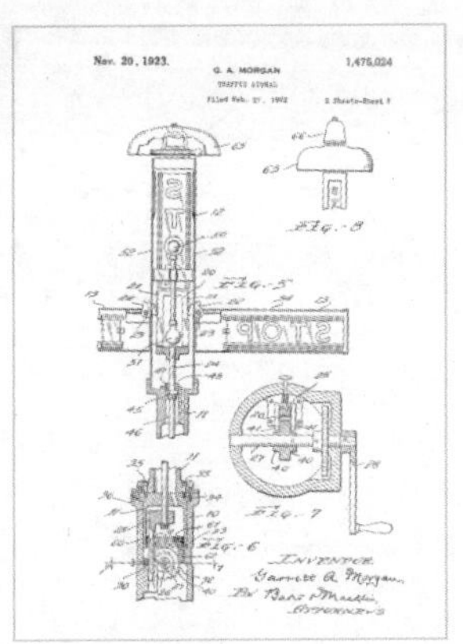

图 11–3

挥命令“停、走、全停”。我们从摩根的专利说明书附图（图 11-2）中可以了解到，摩根信号灯的正面结构图呈十字形，右上、左下、右下方分别是摩根信号灯发出“全停、走、停”三种交通指令时的信号灯正面图。在信号灯立柱下方有一个摇柄，可以通过摇动该摇柄来控制交通信号的发出。当全停指令发出时，行人可以安全的通过道路路口。另外，摩根为了方便夜间指挥交通，还给交通信号灯安装了灯泡，灯泡位于信号灯臂杆的中心，这样就可以将臂杆上的交通指令照亮，方便行人夜间也能看到交通指令。

通过检索专利文献，我们发现后来的发明人从摩根的发明中得到技术启示，并在此基础上进行了新的发明创造。一件是 1954 年，名为“（停车）保护信号设备”的美国 US2669705 号专利，如图 11-4、图 11-5 所示，该发明涉及一种可便携式的设备，用于放在汽车的后面，以提醒道路上行驶的其他车辆注意小心通过，不要撞到停靠车辆或其他人员。这个设备借鉴了摩根发明的 T 形结构设计，两翼和中间一共有 3 个红色的灯，正面图的指令是“注意”“通过”，后视图的指令是“危险”。

Patented Feb. 16, 1954 2,669,705

UNITED STATES PATENT OFFICE

PROTECTIVE SIGNAL DEVICE

图 11-4

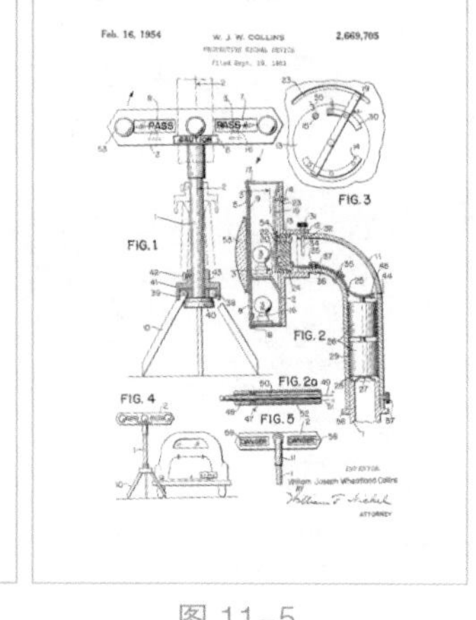

图 11-5

另一件是 1974 年名为“路口交通管理设备”的美国 US3798592

号专利，如图 11-6、图 11-7 所示，该设备的两翼可以根据需要伸展和折起，将其安装在道路交叉路口，通过不同角度的旋转，可达到交通管理指挥的技术效果。

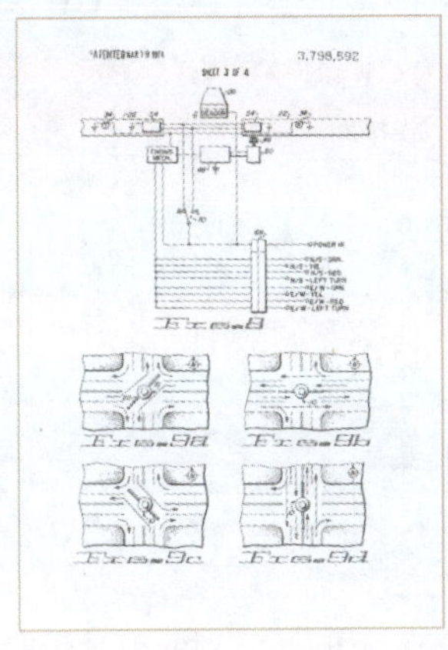

图 11-6

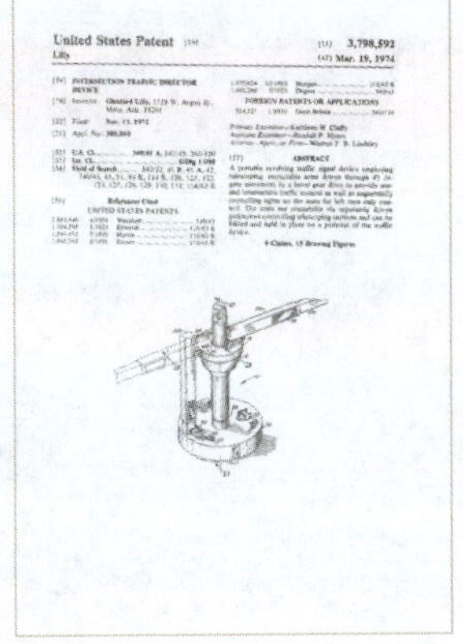

图 11-7

可见，摩根发明的交通信号灯成为后来各种类型交通信号设备的基础。为使自己的发明发挥更大的价值，摩根还将自己的发明申请了加拿大 CA256093 号专利，使其产品在加拿大获得保护。之后，摩根的手摇信号灯交通管理设备被推广至整个北美地区使用，直到被目前全球使用的自动的红色、黄色和绿色的交通信号灯所取代。后来摩根将他的交通信号灯的专利权以 40000 美元的价格出售给美国通用电气公司。1963 年，在摩根去世前不久，他因为交通信号灯的发明获得美国政府的嘉奖。

虽然交通信号灯是摩根最重要的发明，但是并不是他唯一的发明。摩根是一位伟大的发明家和成功的商人，他还发明了很多改变人们生活的物品，例如染发膏、直发器、一种用于手动缝纫机的曲线缝纫附件等。他还成立了摩根精炼公司以推广销售自己发明的产品，让专利转化成为经济价值。

在三角内衣厂大火后，摩根发明了可以保证安全呼吸和防烟雾的面罩，并于 1912 年向美国提交了名为“呼吸装置”的专利申请，

于1914年获得美国US1113675号专利权，如图11-8、图11-9、图11-10所示。通过专利检索可以发现，在美国申请专利保护的同时，摩根还在这款面罩的基础上做了一些技术改良，分别于1912年和1913年向英国和加拿大申请专利保护，并获得英国GB191226312号和加拿大CA155817号专利权。

UNITED STATES PATENT OFFICE.

Patented Oct. 13, 1914.

图 11-8

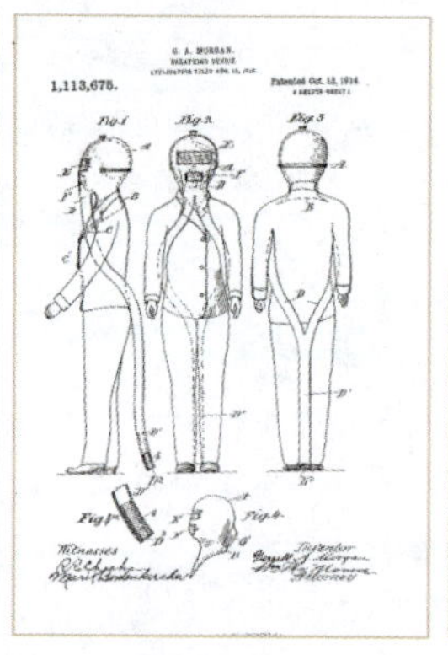

图 11-9

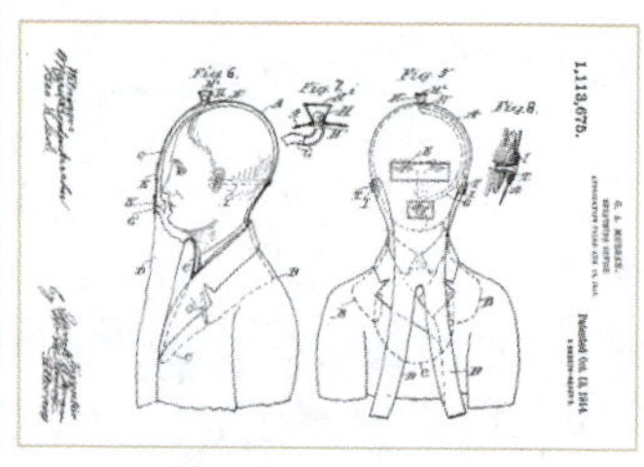

图 11-10

1916年，摩根和一队志愿者戴着自己发明的防毒面罩救下了伊利湖250英尺以下隧道爆炸后的32条生命，这一消息被当时的国家新闻报道出来。之后，摩根的守护神安全设备公司收到了来自全国各地消防部门订购新款防毒面罩的订单。1916年，他早期发明的防毒面具的精致模型在卫生与安全国际博览会获得金奖，而另一个金奖是由国际消防首长协会授予的。美国军队后来在第一次世界大战期间使用了改良后的摩根防毒面罩。

加勒特·摩根于1963年8月27日去世，享年86岁。他的一生是漫长和充实的，他特有的创造性的能量，留给了我们一份奇妙而持久的遗产。

美国 America

1923.US 1475024

交通信号灯

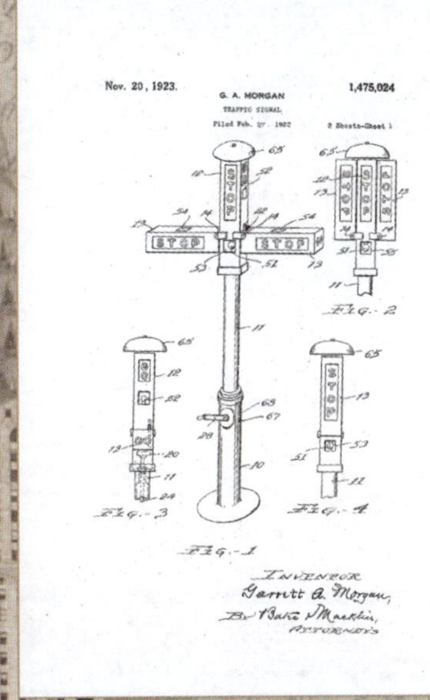

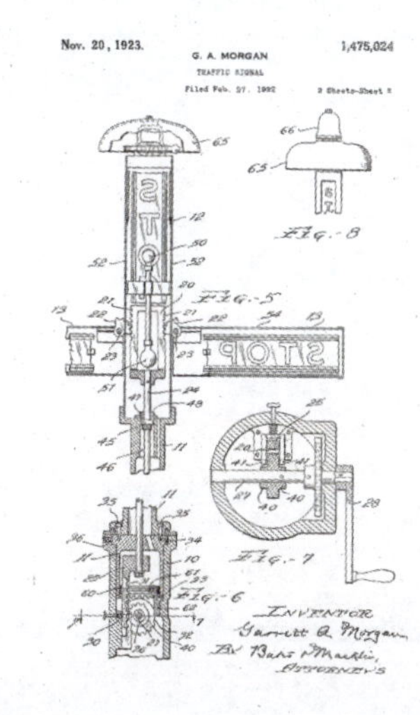

美国 America

Patented Nov. 20, 1923. 1,475,024

UNITED STATES PATENT OFFICE.

GARRETT A. MORGAN, OF CLEVELAND, OHIO.

TRAFFIC SIGNAL.

Application filed February 27, 1922. Serial No. 539,403.

To all whom it may concern:

Be it known that I, GARRETT A. MORGAN, a citizen of the United States, residing at Cleveland, in the county of Cuyahoga and State of Ohio, have invented a certain new and useful Improvement in a Traffic Signal, of which the following is a full, clear, and exact description, reference being had to the accompanying drawings.

This invention relates to traffic signals, and particularly to those which are adapted to be positioned adjacent the intersection of two or more streets and are manually operable for directing the flow of traffic.

One of the objects of my invention is the provision of a visible indicator which is useful in stopping traffic in all directions before the signal to proceed in any one direction is given. This is advantageous in that vehicles which are partly across the intersecting streets are given time to pass the vehicles which are waiting to travel in a transverse direction; thus avoiding accidents which frequently occur by reason of the over-anxiety of the waiting drivers, to start as soon as the signal to proceed is given.

Another object is the provision of a semaphore signal which is useful by night as well as by day and which is arranged to be easily and automatically operable by the traffic director. In addition, my invention contemplates the provision of a signal which may be readily and cheaply manufactured.

To this end, I provide a signal wherein the direction indicating arms are pivotally supported and adapted to be moved vertically for stopping the flow of traffic and then to be revolved and dropped to indicate a right of way to vehicles moving in another direction. The raising and revolving movements of the arms are adapted to be accomplished by the turning of a crank, and suitable mechanism actuated by the crank is provided for automatically indexing the arms to the required position, and for permitting their proper functioning in accordance with the wishes of the traffic director.

The means for accomplishing the above objects will be fully set forth in the following description which relates to the drawings, and the essential characteristics of my invention will be summarized in the claims.

In the drawings, Fig. 1 is a front elevation of a semaphore signal constructed according to my invention; Fig. 2 is an elevation of the direction indicating means showing the arms in vertical position; Fig. 3 is a side elevation of the direction indicating portion when the arms are in horizontal position as shown in Fig. 1; Fig. 4 is a side elevation of the signal when the arms are in vertical position as illustrated in Fig. 2; Fig. 5 is a vertical section through a part of the signalling mechanism; Fig. 6 is a similar section adjacent the operating end of the signal; Fig. 7 is a transverse section taken on the line 7—7 in Fig. 6; and Fig. 8 is a fragmentary view illustrating the signal and alarm bells that may be used with my invention.

My invention is concerned with the provision of a traffic signal which enables a director to control the flow of traffic by stopping the movement thereof in all directions momentarily just prior to allowing the traffic to move in any one direction. Accordingly, in the embodiment shown I have illustrated a standard 10, having a support 11, revolubly mounted therein, which support is provided with direction indicating signals. The signal indicating means may comprise a box-like structure 12 which is rigid with the support 11, and a pair of arms 13 which are pivoted, as at 14, to the opposite sides of the support as illustrated in Fig. 1. The characters shown on the vertical indicator may be the usual "Stop" and "Go" words, as is customary for this class of work.

Each of these arms may be also a box-like structure having a rectangular cross section and having exposed faces provided with direction indicating characters. In the form illustrated the arms are each provided with "Stop" indicating characters on the side and bottom faces, and with "Go" characters on the end faces. The arms are pivoted so that when raised to the position indicated in Fig. 2, the "Go" characters on the indicator 12 are hidden from view, and the "Stop" characters on the bottom of the arms are then visible from the direction in which traffic was moving. The two positions of the arms showing this arrangement are illustrated in Figs. 3 and 4.

2 1,475,024

My invention is adapted to direct traffic by raising the semaphore arms to stop traffic which was moving in one direction and then by revolving the arms and releasing them, thus indicating the right of way for traffic in another direction. The means for raising the arms as shown in Figs. 5 and 6 may be accomplished by lowering a bracket 20 having racks 21 thereon, which actuate segments of a pinion 22, mounted as at 23, on the semaphore arms. This bracket is adapted to be lowered by moving a shaft 24, on which the bracket is mounted, downwardly. This shaft extends entirely through the support 11 and terminates within the standard 10 adjacent another rack 25, which is adapted to be actuated by a pinion 26 mounted on a crank shaft 27. This crank shaft is supported in the standard and is provided with a crank arm 28 which may be rotated by the director.

The segment 26 is shown as having teeth on only a portion of the periphery thereof so that the rack 25 is lowered only during a fractional turn of the crank shaft. Accordingly, to hold the arms in vertical position, I have illustrated a spring operated plunger 30 which is adapted to engage a notch 31 in the rack 25 just before the teeth in the pinion disengage those in the rack. In this way, the arms are held in vertical position whereby traffic is stopped in all directions as indicated in Figs. 2 and 4 respectively.

After the arms are raised and traffic is stopped, then to direct the flow of traffic in a direction different from the flow previous to the raising of the arms, I have illustrated suitable means for revolving the entire signal. This mechanism is adapted to function while the arms are held in vertical position. In Fig. 6, the means for revolving the indicators is shown as a gear 32 rigid with the shaft 27 and provided with teeth on a portion of the periphery thereof for engaging an annular gear 33, which may be integral with a sleeve 34, within the upper portion of the standard 10. This sleeve is shown as being rigidly attached by securing members 35 to the support 11, and since the vertical indicator is rigid with the upper portion of the support, it follows that whenever the gear 33 is rotated, the indicators are revolved.

The degree of turning usually desired is only 90°. Accordingly, I provide an indexing mechanism comprising a yieldable plunger 36 carried by the standard 10 and adapted to engage suitable notches 34. The number of teeth in the gear 32 are so arranged that they just disengage the gear 33 when the plunger 36 engages a succeeding notch. The turning of the indicators is preferably designed to be accomplished only after the arms are moved to vertical position, and this is obtained by spacing the teeth on the gear 32 as illustrated in Fig. 6 so that the gear 33 is not engaged until the pinion 26 disengages the rack 25. Then, while the indicators are being turned, the arms may be released by providing a cam 40 which engages the arms 41, to which the plunger 30 is secured; and thereby releases the rack 25. The weight of the arms 13 causes them to drop to horizontal position.

When the arms are released, the inertia of the moving parts is apt to cause damage to the operating mechanism. Accordingly, I have shown a dashpot for absorbing the shock incident to the fall of the arms and I have shown this dashpot as embodying a piston 45 which is carried by the shaft 24 and is movable within a cylinder 46, forming part of the support 11. Suitable openings 47 may be provided in a closure 48 for regulating the outlet of air from the dashpot cylinder.

To adapt a signal constructed according to my invention for use at night, I have shown two electric lamps 50 and 51 which are mounted within the vertical indicator. These lamps may receive electrical energy either from a battery mounted within the standard 10 (not shown) or from any other suitable source of supply, such as through leads depending from an overhead line. The lamp 50 is positioned adjacent the openings 52 above the point of pivotal connection, while the lamp 51 is adjacent the openings 53 below the point of pivotal connection.

Thus, when the arms are in horizontal position, the lamp 50 illuminates characters on the indicator 12, while the lamp 51 gives an additional warning through openings 53. Then, when the arms are raised, two of the openings 53 hidden by the arms are uncovered, whereupon an illuminated signal is flashed out. This occurs substantially simultaneously with the covering of the "Go" signal shown on the vertical indicator. The "Stop" signal which is flashed when the arms are raised, is augmented by light from the lamp 50 which passes through openings 54 in the top of the arms, and illuminates the characters on the bottom face thereof.

When the source of electrical supply is carried by the standard, the current may be carried to the lamps by means of a collector ring 60 mounted on the sleeve 34, and a brush 61 carried by an arm 62 on the standard 10.

If desired, the signal may be provided with alarm mechanism as illustrated by bells 65 and 66 mounted at the top of the vertical indicators. One of these bells may be an alarm bell, while the other may be a signal bell. For selectively operating these bells, circuit controllers 67 and 68 may be mounted on the standard 10.

1,475,024 3

The operation of a signal constructed according to my invention is as follows:—Assuming that traffic is moving in the direction indicated by the "Go" characters in Fig. 1 and that the director desires to change the right of way, then the crank 28 is turned, whereupon the arms 13 are immediately raised by virtue of the rack and pinion connections actuated by the crank shaft 27. As soon as the arms are raised, the signal which was visible to the moving traffic now changes as shown in Fig. 4, while the signal visible to traffic which was formerly stopped remains the same. Thus, traffic is stopped in all directions. This is accomplished while the signal is moved about ¼ of a revolution. Then as the crank is turned still further, the gear 32 engages the gear 33, whereupon the indicators are revolved while the arms are still held in vertical position. Then while the arms are being revolved, the cams 40 engage the latch arms 41, thereby releasing the plunger 30 and allowing the arms to drop to horizontal position. The dashpot controlled by the piston 45 permits the arms to descend with slow motion so that by the time the crank has been turned a complete revolution, the arms are in substantially horizontal position and are automatically held in the desired direction by the indexing plunger 36. When the arms are lowered, the right of way is changed from the traffic which was flowing in the direction indicated by Fig. 1 to the traffic which was waiting to move in another direction.

When the lights are used, the signals are illuminated to indicate the direction of travel when the arms are in either vertical or horizontal position, and if desired, this may be augmented by bells which are actuated by the circuit controllers adjacent the hand of the director.

From the foregoing description, it will be seen that my invention provides a signal which is operable for enabling a director to control traffic more conveniently than is accomplished by merely revolving semaphore arms in a horizontal plane. A further advantage of my invention is that the movement of the semaphore arms is entirely automatic whereby the operator may devote his attention to the requirements of traffic and still accomplish his purpose without manually signalling, moving traffic to stop, until the direction of travel is changed.

Having thus described my invention, I claim:—

1. In a traffic signal the combination with a vertically disposed indicator, of semaphore arms pivoted thereto and revoluble thereon, said arms being disposed on opposite sides of the indicator, means for raising and lowering the arms and crank mechanism for controlling said means, said mechanism being so arranged that the arms are raised during the initial turning of the crank and are then revolved upon continued turning of the crank.

2. A traffic signal comprising in combination, a standard, a vertical indicator carried thereby, a semaphore arm pivoted to the indicator, a crank carried by the standard, means associated with the crank for raising the arm and subsequently revolving the indicator, and other means for permitting the arm to be lowered in a position at right angles to the former position when the crank has been turned further.

3. In a traffic signal, the combination with a standard, of a revoluble indicating member carried thereby, semaphore arms disposed on opposite sides of said member and pivoted thereto, a crank shaft projecting laterally through the standard, mechanism within the standard and member for raising the arms to vertical position subsequently revolving the member and then lowering the arms, whereby traffic is momentarily stopped in all directions and then subsequently permitted to flow in another direction, said mechanism being so arranged that the signal is actuated to automatically stop traffic in all directions on each operation of the crank shaft.

4. In combination, a standard, a revoluble member projecting upwardly therefrom, semaphore arms pivoted to the indicator and disposed on opposite sides thereof, a shaft extending vertically through the member, rack and pinion means at each end thereof for raising said arms, a crank actuated member carried by the standard for controlling the rack and pinion means, a latch carried by the standard for retaining the shaft in lowermost position, and a cam controlled by the crank shaft for releasing the latch, whereby the arms are permitted to drop to horizontal position.

5. In combination, a standard, a revoluble member carried thereby, a vertically disposed indicator secured to the upper end of the member, semaphore arms pivotally disposed on opposite sides of the indicator, mechanism extending within the indicator, a standard for raising the arms, a crank actuated mechanism carried by the standard, said last mentioned mechanism being so arranged that the arms are raised during the initial turning of the crank shaft and then revolved upon continued turning of the shaft.

6. In combination, a standard, a hollow member projecting upwardly therefrom, an indicator rigid with the upper end of said standard, semaphore arms pivoted to the indicator, mechanism controlled from the standard for actuating the arms to move them in a vertical plane and for revolving them with the indicator to a new position, means associated with the standard for indexing the indicator to the proper position, means associated with said standard for retaining the arms in a vertical position during a portion of the turning operation and for releasing the arms when moved to a new position, and a dash pot associated with said member for cushioning the shock incident to the fall of said arms.

4 1,475,024

7. In a traffic signal, the combination with a vertical indicator having openings in opposite sides thereof, of a semaphore arm pivoted thereto below the openings, said arm having openings in the upper wall thereof which are adapted to register with openings in the indicator when the arm is raised to vertical position, a lamp within the indicator adjacent the openings therein, and means for operating said arm.

In testimony whereof, I hereunto affix my signature.

GARRETT A. MORGAN.

创可贴的发明
Band-Aid

埃尔·迪克森
Earle Dickson

创可贴，日常生活中常用的一种外科用药，可方便、简捷地处理伤口。创可贴由埃尔·迪克森于1920年发明。

埃尔·迪克森在帮助妻子包扎伤口时，萌发了制作一种自助包扎伤口绷带的想法。他考虑到，如果把纱布和绷带做在一起，就能用一只手包扎伤口。经过反复试验，他采用了一种粗硬纱布，在上面涂上胶，然后把另一条纱布折成纱布垫，放在绷带的中间，制作出第一款创可贴。

迪克森发明的创可贴由美国强生公司生产后投入市场，获得巨大成功。该发明于1926年获得美国专利。

US 1612267

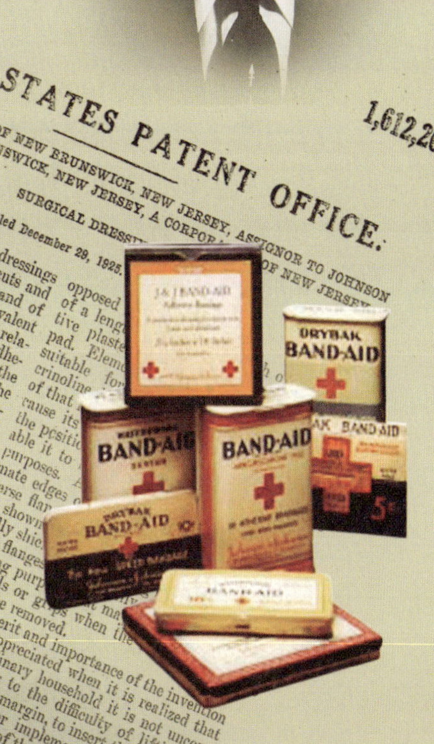

www.sipo.gov.cn

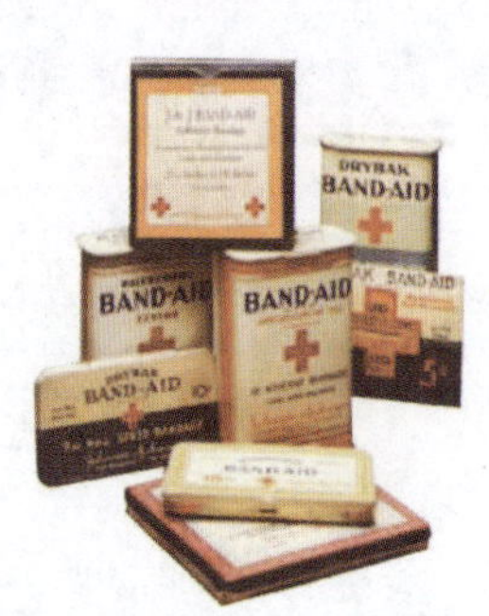

第十二篇 创可贴

作为现代人最常用的一种外科止血胶布，创可贴已经成为每个家庭的生活必备品。有资料显示，全世界每年要用掉将近10亿个创可贴，难怪有人将它列为20世纪影响生活的十大发明之一。而看似不起眼的小创可贴，它的发明背后还有一个感人的故事。

20世纪初，在美国西部的一个小城，刚刚结婚的迪克森太太对烹调毫无经验，常常在厨房里切到手或烫到自己。那时，丈夫埃尔·迪克森（Earle Dickson，1892–1961年）正在一家生产外科手术绷带的公司里工作，他每次称赞太太厨艺进步的时候，都要为妻子的手指担心。每天回到家第一件事，迪克森不是吃饭，而是先帮妻子重新包扎伤口。为了妻子，迪克森决定发明一种绷带，让妻子在受伤而无人帮忙时能自己包扎伤口。

迪克森开始做实验，他考虑到，如果把纱布和绷带放到一起，就能用一只手来包扎伤口。于是，迪克森拿了一条纱布摆在桌子上，在上面涂上胶，然后把另一条纱布折成纱布垫，放到绷带的中间。但做这种绷带的粘胶暴露在空气中时间长了就会干。迪克森试着用许多不同的布料盖在胶带上面，期望找到一种在需要时不难揭下来的材料。经过多次试验之后，他发现，一种粗硬纱布能很好地完成这个任务。经过不断地改近，最初的“创可贴”诞生了。

1925年，迪克森向美国提交了名为“外科敷料”（Surgical

dressing）的专利申请，并于1926年获得美国US1612267号专利权，如图12-1、图12-2所示。迪克森发明的创可贴绷带给他带来了好运。他所在的公司主管将它命名为Band-Aid，Band指的是绷带，而Aid是帮助急救的意思，也就是“邦迪”。公司就把“邦迪”作为急救绷带产品的名称。随后，这种创可贴行销世界。

Patented Dec. 28, 1926. 1,612,267

UNITED STATES PATENT OFFICE.

图12-1

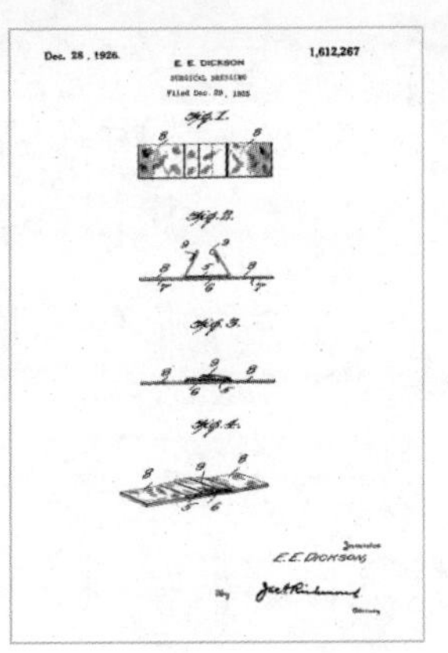

图12-2

从专利文献中我们能够直观地看到外科敷料的结构。该发明技术方案是一种外科敷料，包含一条胶带，一块柔软的纱布，为保证纱布的无菌性，其与胶带并不等宽，而是相对略窄，另有两条绷带黏附在暴露的胶带上，这两条绷带从相对方向互补，所述的面对面位置的绷带和纱布从接近尾端形成互锁的结构使它们保持相对于平坦的覆盖关系。

那么，迪克森的发明对当时的技术进步起到了哪些影响呢？我们通过专利检索可以发现，此项专利构思启迪了后来发明者的发明创造。翻看这些后来发明者的专利文献会发现，从技术领域来看他们的发明与迪克森的发明关联并不大，但是从发明的技术要点上便可发现，这些技术方案都在一定程度上受到迪克森专利技术的启示。

例如，1955年名为“一种印刷方法”的美国US2699103号专利，如图12-3、图12-4所示，它将手写体的英文字母分别印在类似玻璃纸的媒介上，然后粘贴在黏性胶带上，这样当需要印刷时可以撕下所

需要的字母排列成单词使用。这一发明被应用于 20 世纪四五十年代的广告领域，能够满足当时广告领域艺术字体印刷的需要。这一发明的技术方案就应用了迪克森创可贴发明中可揭开使用的绷带的技术特征，但是同一技术特征应用于不同的技术领域就产生了不同的技术效果，也解决了不同的技术问题，从而诞生了一项新的发明。

United States Patent Office

2,699,103
Patented Jan. 11, 1955

1

2

图 12-3

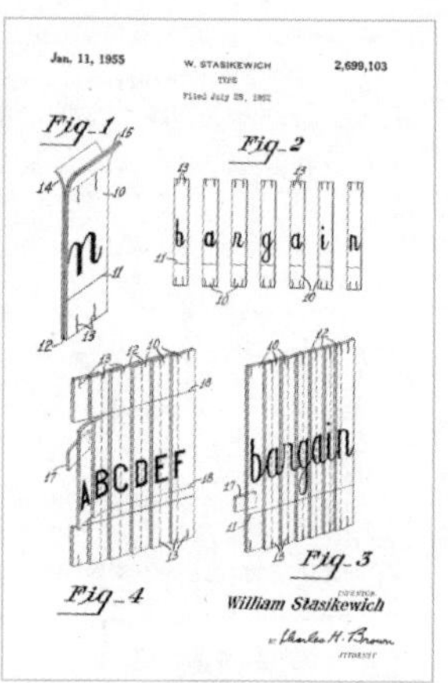

图 12-4

在专利文献中，可以找到许多相关的发明，最具代表性的一件发明是止血黏性绷带。这一发明 1985 年 6 月在美国申请了专利，并于 1986 年 10 月获得美国 US4616644 号专利权，如图 12-5、图 12-6 所示。通过阅读该发明的技术方案可知，该发明是用于小割伤与创伤的止血胶粘绷带，由一个内科或外科胶粘绷带组成，包括一个胶带衬底，其上贴着一个吸收垫，垫上覆盖着一个多孔的塑料膜型非黏性的伤口防粘盖，伤口防粘盖上涂有很薄一层由分子量大于约 60 万道尔顿的聚氧乙烯构成的止血剂，该绷带止住小割伤的出血快，若所用的聚氧乙烯的分子量更高，则达到同样止血效果的需用量还要少。

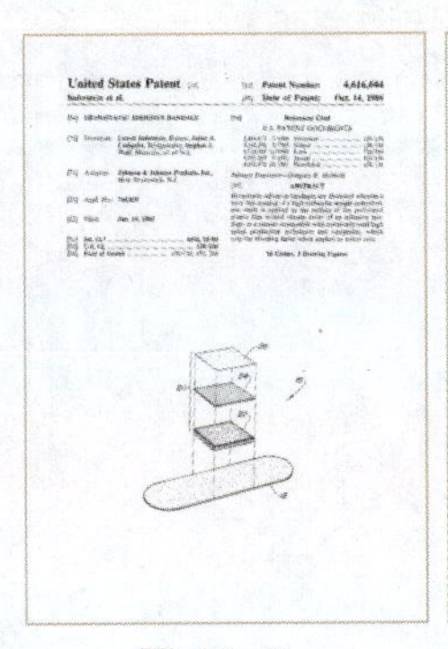
图 12-5

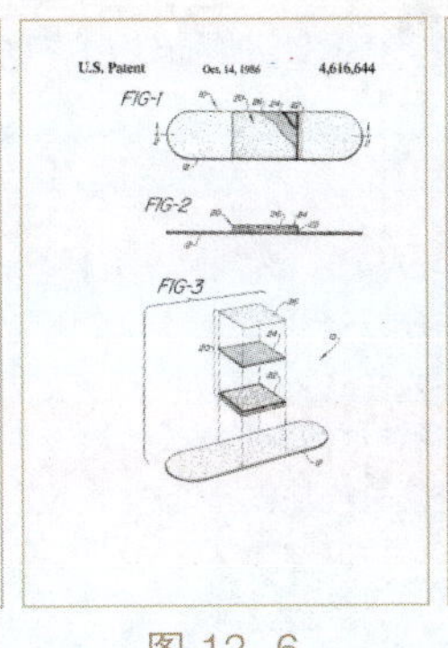
图 12-6

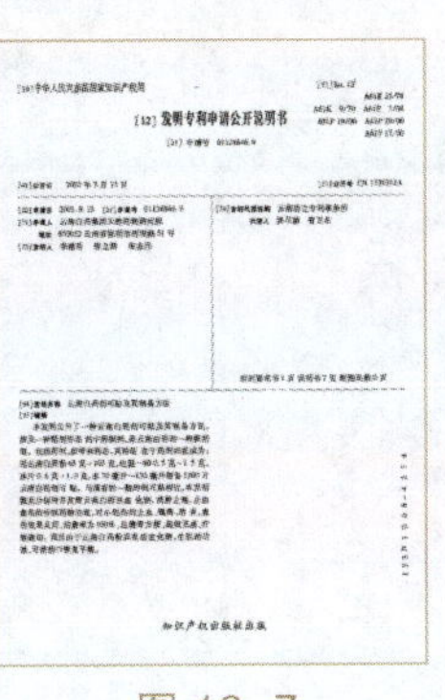
图 12-7

2004年获权的我国CN01128846号发明专利，如图12-7所示，将有着100多年历史、被称作“疗伤圣药”的云南白药经过特殊工艺与药芯、胶布进行完美结合，制造出云南白药创可贴。它与市售的其他“创可贴”相比，能充分保持并发挥云南白药活血化瘀、消肿止痛、止血愈伤的传统药物功效，对小创伤的止血、镇痛、消炎、愈伤，效果良好，治愈率为100%，且携带方便、起效迅速、疗效确切，而且由于云南白药粉具有活血化瘀、生肌的功效，可使伤口恢复平整。2009年该专利获得第11届中国专利奖优秀奖。

如今，这种简单实用、能够包扎小创伤的物品已经成为家庭生活的必备品。它的功能也在不断拓展，能预防晕车，甚至可以修补帐篷或睡袋。可以说，这项小小的发明给人类带来了极大的方便。

美国 America

1926. US 1612267

——外科敷料

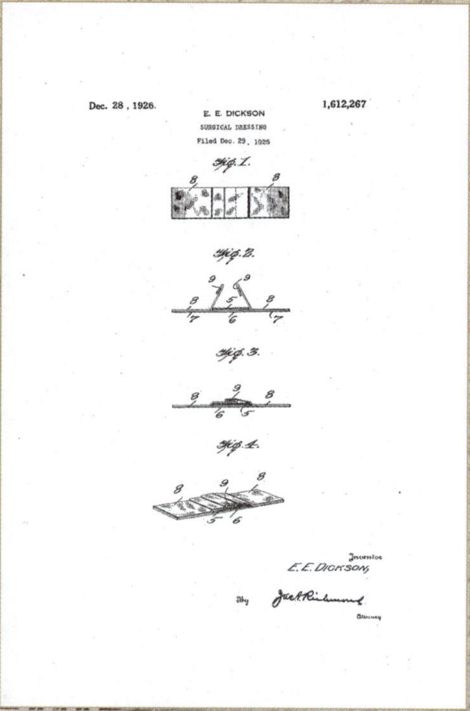

Dec. 28, 1926. E. E. DICKSON 1,612,267

SURGICAL DRESSING

Filed Dec. 29, 1925

Fig. 1.

Fig. 2.

Fig. 3.

Fig. 4.

Inventor

E. E. DICKSON,

By

Attorney

Patented Dec. 28, 1926. 1,612,267

UNITED STATES PATENT OFFICE.

ENSIGN EARLE DICKSON, OF NEW BRUNSWICK, NEW JERSEY, ASSIGNOR TO JOHNSON & JOHNSON, OF NEW BRUNSWICK, NEW JERSEY, A CORPORATION OF NEW JERSEY.

SURGICAL DRESSING.

Application filed December 29, 1925. Serial No. 78,131.

My invention relates to surgical dressings mainly intended for application to cuts and other injuries of a minor character and of the type embodying a gauze or equivalent pad or compress which is positioned relative to the part to be dressed by an adhesive strip or plaster. A usual adjunct of the dressing is a face cloth which serves the dual office of shielding the pad until required for service so as to keep it clean and sterile and of preventing promiscuous sticking of the plaster. When emergency arises for use of the dressing the facing cloth is of course removed and this is accomplished generally with difficulty and with casual contact with the pad, resulting in infection, thus nullifying the precautionary steps taken in the manufacture and packing of the article.

The nature of my invention consists in providing a partible interlocking facing element which will efficiently perform the dual office stated and yet, in emergency, be detached readily and without liability of impairing the sterility of the pad.

My invention is illustrated in the accompanying drawing, wherein—

Figure 1 is a plan view of a faced band dressing.

Fig. 2 is an edge view with the facing components in open or unlocked relation.

Fig. 3 is an edge view with the facing elements interlocked.

Fig. 4 is a perspective view, likewise showing the facing elements interlocked.

In practising my invention, a pad or compress 5 of gauze or other appropriate material is applied to the adhesive side of a strip or band 6 of adhesive plaster in such way as to leave margins 7 of the latter free for service in securing the pad to place with relation to the part to be dressed. In the drawing the pad is arranged medially of the plaster and, obviously, it is retained by the adhesive element thereof.

As a coverage for the pad and the adhesive side of the plaster I provide endwise opposed facing elements 8 each of which is of a length and width to cover the respective plaster margins and also overlie the pad. Elements 8 may be made of any fabric suitable for the purpose. For example, crinoline has given satisfaction. A fabric of that nature has sufficient body stiffness to cause its free ends to assume substantially the position shown in Fig. 2 and also to enable it to be edge formed for interlocking purposes. Accordingly I provide the proximate edges of the facing elements with reverse flanges 9 in order to interlock them, as shown in Figs. 3 and 4, so as to effectually shield the underlying pad or compress. The flanges are not only useful for interlocking purposes but may also be availed of as pulls or grips when the facing elements are to be removed.

The merit and importance of the invention will be appreciated when it is realized that in the ordinary household it is not uncommon, owing to the difficulty of lifting the facing at its margin, to insert the finger or a knife or other implement below the unsecured portion of the facing thus contaminating the pad and making for infection. With the tabs or pulls afforded by the interlocking facing elements of my invention that menace is obviated.

Having described my invention I claim:—

A surgical dressing embodying a strip of adhesive tape, a soft dressing offset with relation to the tape, and means to insure sterility of the dressing comprising complemental facing strips in adhering relation to exposed portions of the tape and of body stiffness sufficient to spring them away from the dressing, said facing strips having their proximate ends formed with interlocking elements whereby they are retained in flat coverage relation with respect to the dressing.

In testimony whereof I affix my signature.

ENSIGN EARLE DICKSON.

伸缩雨伞的发明

Telescopic Umbrella

汉斯·霍普特
Hans Haupt

18世纪，巴黎人用油布做伞面，制成了真正的雨伞。1874年，英国人霍克斯设计出了弧形钢质伞骨，也就是使用至今的长柄雨伞。而直到1920年德国人汉斯·霍普特对伞具进行了改进，才发明了可折叠雨伞。

该项发明1934年获得德国专利（DE606015），由于轻便实用，在市场上广受欢迎，因此霍普特成立了公司专门生产可折叠雨伞，后来发展为著名的Knirps伞具品牌，Knirps在德语中有“小家伙”和“弹出”的含义。

Knirps品牌随即在全世界范围内引发了一场雨伞的大变革。1936年，基于折叠雨伞发明构思的更先进的发明——自动伸缩弹出雨伞出现在德国市场上。

DE 606015

DEUTSCHES REICH

AUSGEGEBEN AM 23. NOVEMBER 1934

REICHSPATENTAMT

PATENTSCHRIFT

№ 606015

KLASSE 33a GRUPPE 9

www.sipo.gov.cn

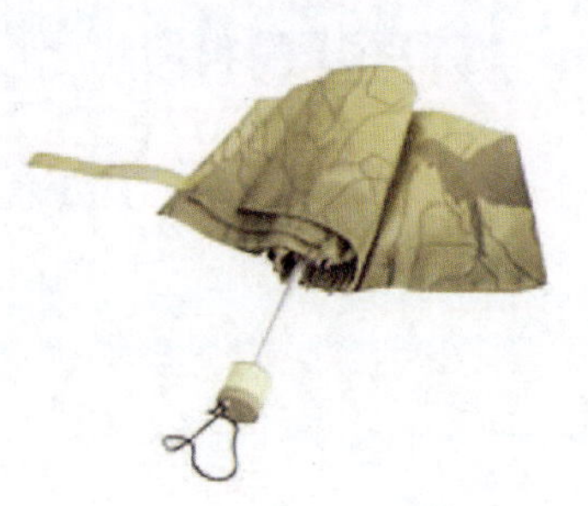

第十三篇 伸缩雨伞

伞在人类生活中的历史可以追溯到4000年前，古埃及、古希腊、古中国的文明中都存在关于伞的记载。伞的英文单词“umbrella”来源于拉丁语“umbra”，是“阴影”的意思，因此伞最早是用于遮阳。中国人利用油纸制成了世界上最早的防水伞。

欧洲到17世纪才逐步普及伞的使用，而且是从中国传入的，在这之前欧洲人都使用斗篷来遮风挡雨。伞最初被视为女性饰品。早期的伞是不能收合的，1710年法国人发明了欧洲第一把能收合的伞，又用油布做伞面，制成了真正的雨伞。此后人们一直在对伞进行改进，涉及伞面的材质、伞骨的材质、伞骨的结构、易携带性等。1852年法国人发明了能自动打开的伞。1874年英国人设计出了弧形钢制伞骨，也就是使用至今的长柄雨伞。长柄雨伞是英国绅士的象征，但是不便于携带。

1920年，德国工程师汉斯·霍普特（Hans Haupt）对伞具进行了改进，发明了可折叠雨伞，被认为是世界上最早的折叠雨伞。由于腿在战争中受伤，汉斯行走时需要使用手杖。但同时携带手杖和长柄伞给他的出行带来了不便，于是他发明了具有伸缩式框架的伞，收缩后几乎能放到口袋里，所以也被称为口袋伞（pocket umbrella）。

1930年，汉斯对他的发明在德国提交了专利申请，1934年获得了名为“折叠伞”的德国DE606015号专利权，如图13-1所示。通

过检索专利文献，我们发现，1931 年，汉斯就这个发明在美国也提出了申请并于 1933 年获得美国 US1902774 号专利权，如图 13-1、图 13-2、图 13-3 所示。这篇美国专利文献帮助我们解决了阅读德文文献的困难，同时通过专利文献中的附图，我们能够直观地看到这项发明中伸缩雨伞的结构。

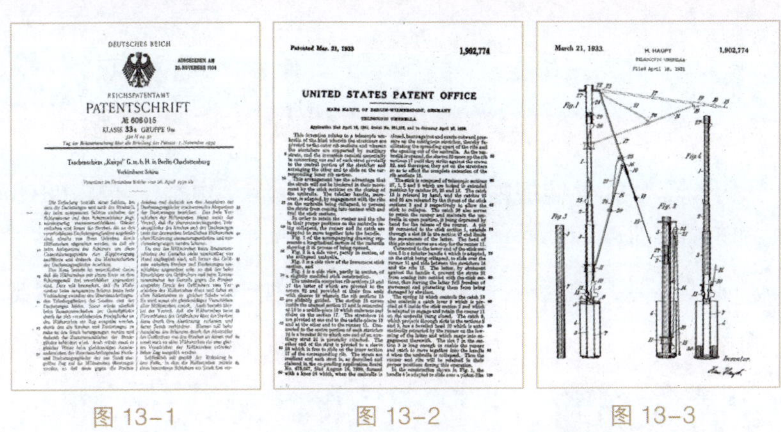

DEUTSCHES REICH

REICHSPATENTAMT

PATENTSCHRIFT

№ 606015

KLASSE 33a GRUPPE 9

Patented Mar. 21, 1933　1,902,774

UNITED STATES PATENT OFFICE

March 21, 1933.　H. HAUPT　1,902,774

图 13-1　　图 13-2　　图 13-3

如专利文献所示，在这项发明中，汉斯将伞骨分为两部分，伞柄也被设计成多部分，伞骨和伞柄均可伸缩；汉斯为伞骨支架添加了辅助支撑杆——这正是在伞撑开过程中带动伞骨伸缩的关键部件。

那么，汉斯发明的伸缩雨伞和同时期的其他雨伞有什么区别呢？他又是如何一步步改进最终形成这项发明的呢？汉斯之后，伸缩伞又有哪些发展变化呢？通过检索专利文献同样能够帮助我们找到答案。

在汉斯的发明前，专利文献中已经出现了与辅助支架类似的结构。如，1897 年在德国申请的一项关于自动弹出雨伞的德国 DE100186 号专利，如图 13-4 所示，其伞骨和伞柄之间有伞骨支架和辅助弹簧连接的设计。在这项发明中，辅助弹簧解决的是使伞自动弹开的问题。而汉斯将辅助弹簧改为了辅助支架，将其位置移动到了伞骨和伞骨支架之间，解决了伞骨伸缩的问题。辅助支架也是汉斯的发明与同时期其他雨伞（特别是也具备伸缩功能的雨伞）的显著区别。

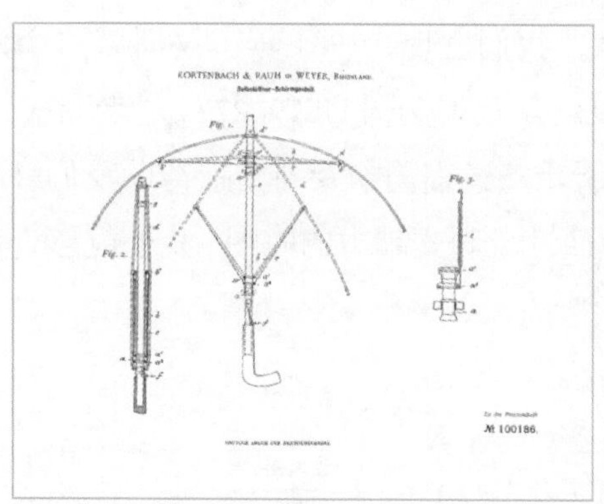

图 13-4

汉斯的首个关于伸缩雨伞的专利是在 1914 年申请的英国 GB2065 号专利，如图 13-5 所示。可以看到，在这项发明中他已经将伞骨和伞柄设计成多段。随后，他在伞骨和伞骨支架之间增加了辅助支架，使伸缩雨伞初具雏形。在不断改进发明的同时，汉斯也持续通过申请专利保护他的发明创造。1914 ~ 1957 年，汉斯申请了 30 多项关于伸缩雨伞的专利，这些专利主要集中在英国、美国，还在加拿大和爱尔兰各申请了一项专利。这也体现了汉斯有意将他发明的雨伞销售到这几个国家。

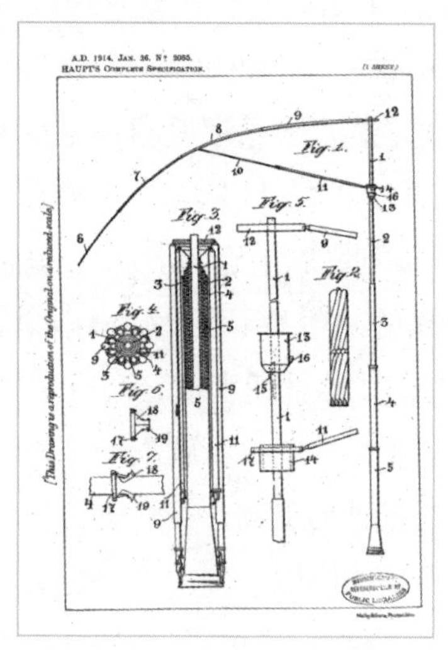

图 13-5

后来的发明人在汉斯的发明基础上对伸缩雨伞进行了改进。后来的这些专利申请中，有的发明人将辅助支架的位置从伞骨和伞骨支架之间改变为伞骨支架与伞柄之间，如 1949 年一项名为“伞的骨架结构”的德国 DE877186 号专利，如图 13-6 所示。伸缩雨伞发展到今天，其结构有了更多的变化和更加复杂的设计。如 1993 年一项名为“一种三折伞的伞骨”的中国 CN2162120 号实用新型专利，如图 13-7 所示，伞的辅助支架被放置到了伞骨支架和伞柄之间，伞骨的收缩方式也由伸缩变成了折叠。

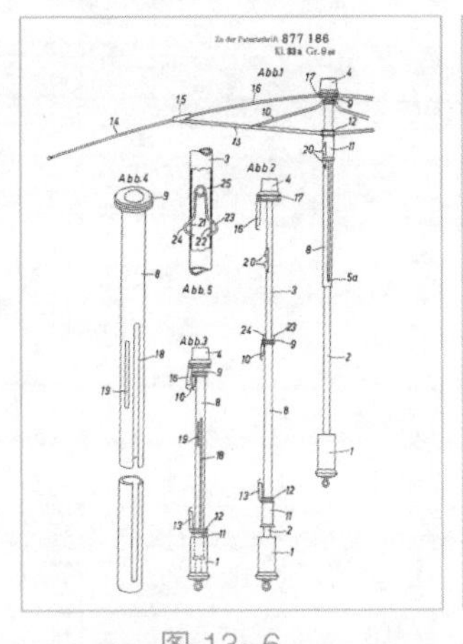

图 13-6

[12] 实用新型专利说明书
[21] ZL 专利号 93213806.3
A45B 25/02
(BJ)第 1452 号

图 13-7

汉斯的发明不但为我们带来了“口袋伞”这款经典的雨伞，还促使了一个世界著名的雨伞品牌的诞生。汉斯发明的口袋伞由于轻便实用，在市场上广受欢迎，因此他成立了公司专门生产伸缩雨伞，逐渐发展为后来的“克尼普斯”（Knirps）伞具品牌，“Knirps”在德语中有“小家伙”和“弹出”的含义，也就是著名的德国红点伞。今天，“Knirps”已经成为世界最为著名的雨伞品牌之一，其以优良的技术、先进的材质、过硬的品质而著称。2003 年 Knirps 获得了世界知名的红点设计奖（Reddot Award）。在德国，“Knirps”是家喻户晓的品牌，“下雨天”更被称为“Knirps Weather”。“Knirps”今天的成绩来源于汉斯的发明，更与知识产权对品牌的保护与提升密不可分。

1934.DE 606015

——折叠伞

DEUTSCHES REICH

AUSGEGEBEN AM 23. NOVEMBER 1934

REICHSPATENTAMT

PATENTSCHRIFT

№ 606015

KLASSE 33a GRUPPE 9

Taschenschirm „Knirps" G. m. b. H. in Berlin-Charlottenburg

Verkürzbarer Schirm

Zu der Patentschrift 606015

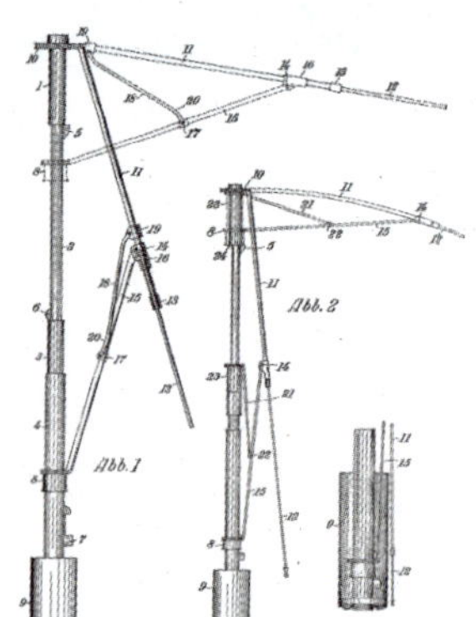

圆珠笔的发明 Ball Pen

拉兹罗·约瑟夫·比克
Laszlo Josef Biro

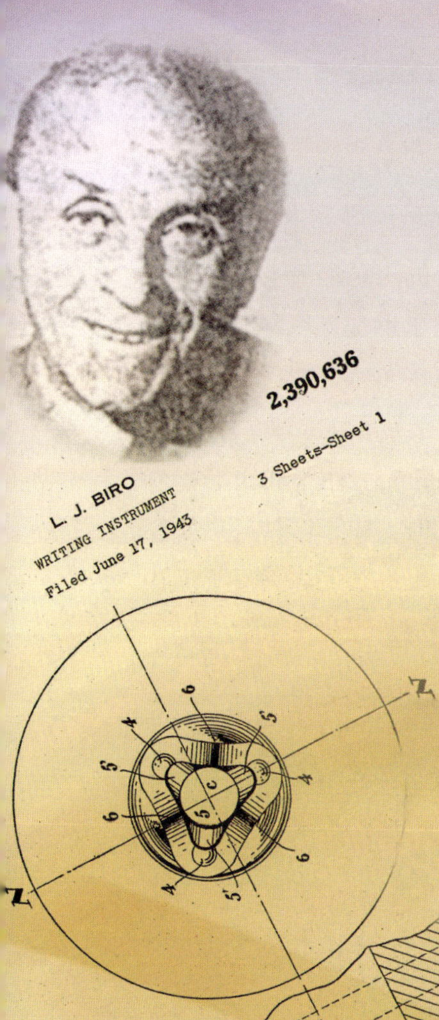

2,390,636
3 Sheets-Sheet 1
L. J. BIRO
WRITING INSTRUMENT
Filed June 17, 1943

圆珠笔是利用钢珠的旋转把油墨写到纸上的一种书写工具，不渗漏、不受气候影响，并且书写时间较长。它由匈牙利人拉兹罗·约瑟夫·比克发明。

1943 年，匈牙利的一位编辑拉兹罗·约瑟夫·比克发现，印刷油墨干得很快而且不会用脏稿纸，于是他决定制造一种使用同样油墨的笔。可是这种黏稠的墨水无法从一般的笔尖流出来。在身为化学家的兄弟乔治的帮助下，他们设计出一种利用毛细作用把油墨送至圆珠的笔，于是，世界上第一支圆珠笔诞生了。

1945 年比克获得了圆珠笔的专利（US2390636）。后来，比克将这项发明提供给英国皇家空军。不久，英国的一家飞机制造厂就推出了首批商业化的圆珠笔。第二次世界大战后，由于开发出了低成本的制造方法，圆珠笔很快在全世界流行起来。

US 2390636

www.sipo.gov.cn

第十四篇
圆珠笔

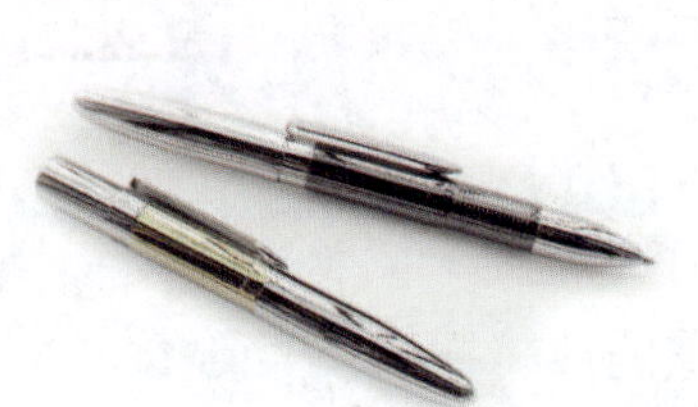

圆珠笔是利用钢珠的旋转把油墨写到纸上的一种书写工具，书写的时候，笔芯里存储的墨水通过毛细作用润滑了钢珠，均匀地渗出墨水，达到书写的目的。圆珠笔不渗漏、不受气候影响，并且书写时间较长，已成为近数十年来风行世界的一种书写工具。

在圆珠笔问世以前，人们常用的书写工具是自来水笔和铅笔。然而，自来水笔存水少，不适合长时间书写，同时，由于这种自来水笔的墨水晾干需要一定时间，使用者经常会在手上、纸上沾满墨渍，很难清洗。而铅笔的质量也不稳定，字迹模糊不清，还容易擦除，许多文件不适合使用铅笔书写。

圆珠笔的原理最早可以追溯到 1888 年。这一年，约翰·劳德（John J. Loud）发明了一种笔尖带有旋转圆珠、能从墨水囊中不断获得墨水的钢笔。尽管这种钢笔也有漏水的缺点，但在皮革上写字要比自来水笔好用得多。之后，他对此项发明申请了加拿大 CA32643 号专利。遗憾的是，他没有充分利用该专利技术，这一专利并未商业化，从而也无人知晓。

现在广泛使用的圆珠笔诞生于 1938 年，是由一名匈牙利的编辑拉兹罗·约瑟夫·比克（Laszlo Josef Biro）发明的。那时，作为记者的比克常常跟同事抱怨他们所使用的旧式墨水笔，因为这种笔漏水，工作时手上、纸上总会沾满墨渍，很难清洗。20 世纪 30 年代后期，

当比克在一家报社看到速干墨水时，他产生了一个新想法：如果在自来水笔的胆管里装上像油墨一样的特别墨水，那么写的字不也能很快干燥吗？1938 年 7 月，比克制作出第一支名为“油溶笔”的圆珠笔。1938 ~ 1939 年，他先后向多个国家递交专利申请并获得了法国 FR853022 号、英国 GB512218 号、美国 US2265055 号三个国家的专利权。此后，他不断尝试在“油溶笔”基础上改进圆珠笔的技术方案。

“二战”前夕，比克的祖国受到德国法西斯的胁迫，1940 年为了躲避纳粹的迫害，比克移民到阿根廷。但是，比克并没有间断对圆珠笔的研究，仍以极大的热情投入到圆珠笔技术方案的改进中。1943 年 6 月，比克在阿根廷向美国递交了专利申请，并于 1945 年 12 月获得名为“书写工具”的美国 US2390636 号专利权。这项专利被认为是比克获得的圆珠笔系列专利中最有代表性的专利。该发明根据油性渗透原理，采用油溶性颜料作为墨水，在大气压力与油墨重力的双重作用下，笔芯中的油墨通过油管流向笔头的小钢珠座里，然后油墨黏附在球珠上。书写时，黏附在球珠上的油墨随着小钢珠在书写面上的滚动而黏附在书写面上，形成字迹，即达到书写的目的。这项技术主要改进了自来水笔的吸管，并将笔尖改为能任意转动的小圆珠，从而使墨水平滑地写在书写面上，如图 14-1、图 14-2、图 14-3 所示。

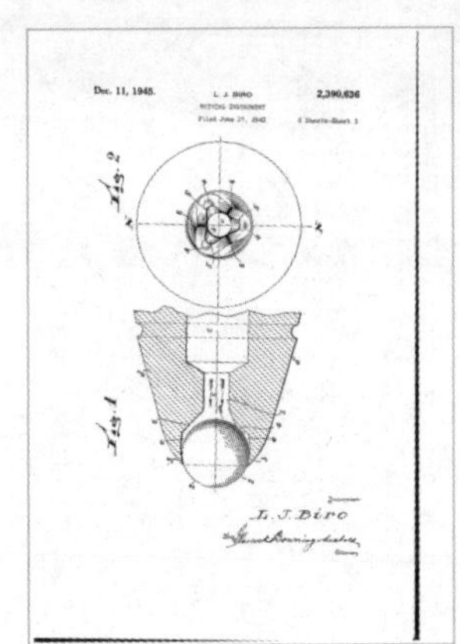

图 14-1

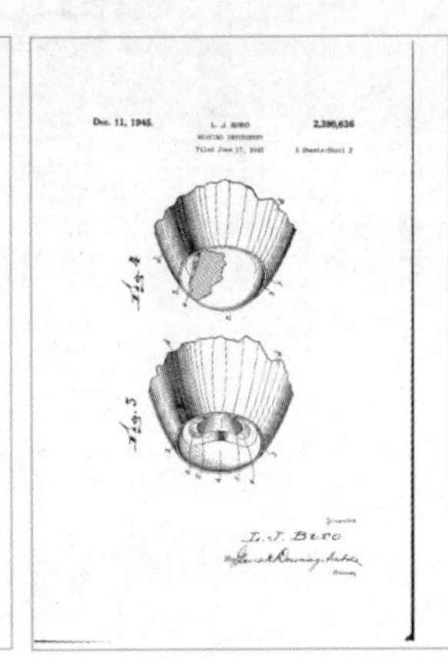

图 14-2

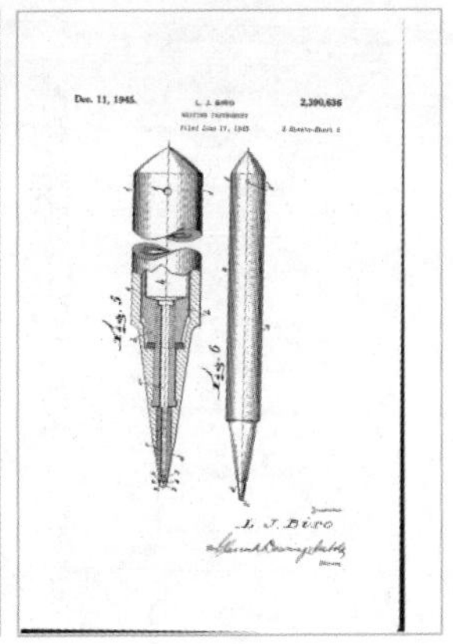

图 14-3

此后，比克依然就圆珠笔的墨水储存与流量控制等细节不断钻研、改进。他首先将笔芯设计为螺旋形状，并通过控制气压来控制墨水的流量，如图 14-4、图 14-5 所示。此后不久，为提升一支笔的总墨量，他将笔芯又改进为回旋形状，如图 14-6、图 14-7 所示。整个过程中，每一个阶段性的成果都更加鼓舞比克继续研究下去，为保护自己的智慧结晶，他申请并获得了多项美国专利（US2397229，US2413904，US2416896，US2400679）。

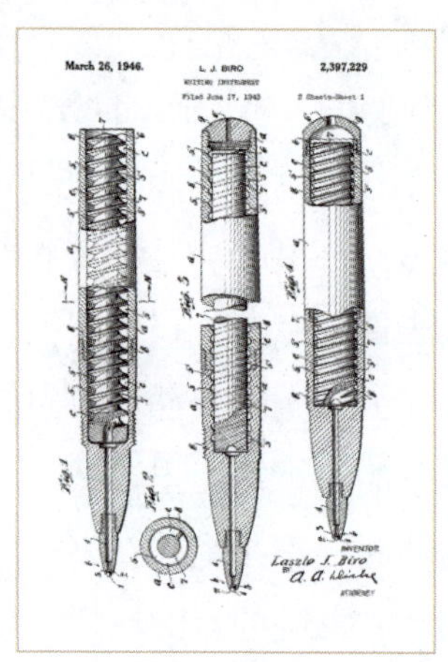

图 14-4

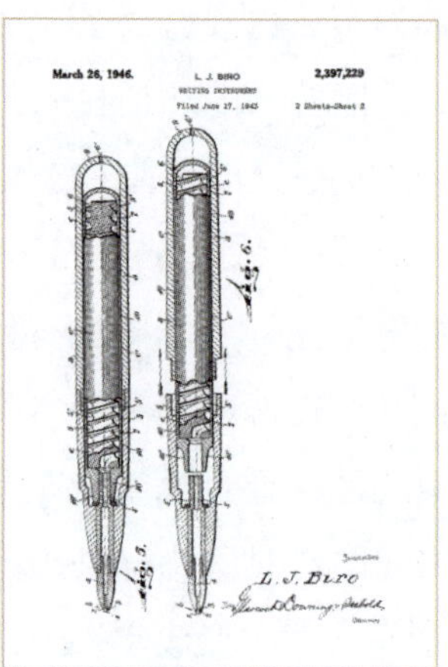

图 14-5

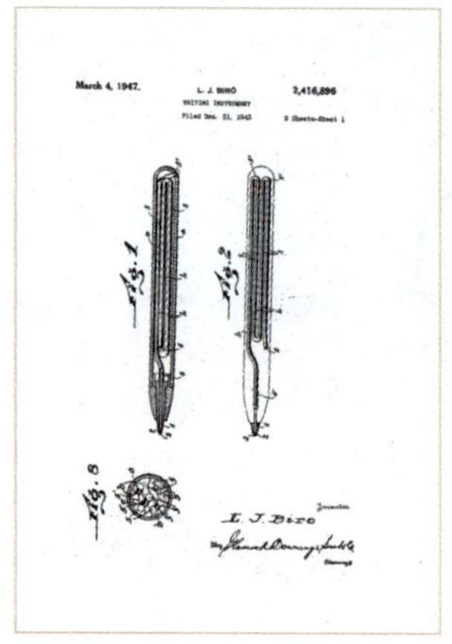

图 14-6

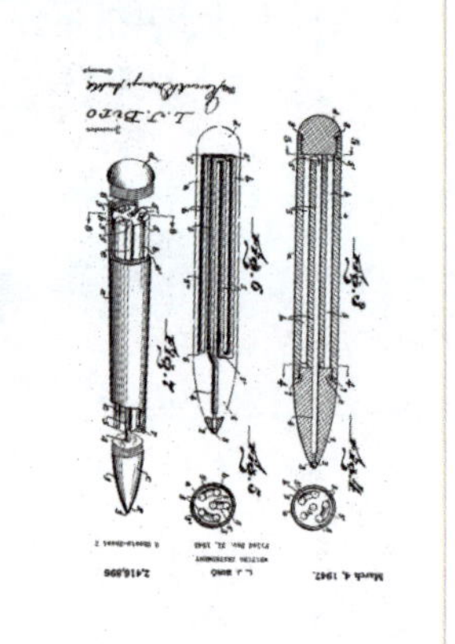

图 14-7

经过几年时间的研究试验，比克终于完成了可以投入商业使用的比较完善的圆珠笔技术方案。比克在一位金融家的帮助下，研制出生

产圆珠笔的机器。1943 年，比克在阿根廷办的圆珠笔小厂生产出第一批产品。为了纪念发明人比克，人们将圆珠笔称为“比克笔”。

时值“二战”，英国皇家空军急需一种在高空战斗机上不漏水的笔，因此，英国政府出资买下了这个专利使用权。不久，英国的一家飞机制造厂就推出了首批商业化的圆珠笔。从此，“比克笔”备受人们瞩目。

最初推向市场的圆珠笔要价高达 10 美元，这在当时是很贵的。因为，笔尖上直径约 0.1 厘米的小钢珠成本很高。小钢珠是由铬和钢的合金制成，非常耐压、耐磨，它也是圆珠笔最大的特色所在。

虽然如此，很多厂商依然看到了潜在的商机。为抢占市场，他们纷纷着手进行技术上的改进。这其中就包括全球著名的制笔公司——派克钢笔公司。

1945 年 7 月，为降低圆珠笔的使用成本，减少价格昂贵的小钢珠的使用，派克公司在其发明的美国 US2452504 号专利中公开了一项技术，对圆珠笔笔芯进行了改进，增大了圆珠笔笔芯的容量，并使容纳油墨的笔芯部分可以通过螺旋方式与笔头部分进行连接，在油墨使用完毕后，可以更换存储油墨的笔芯，从而减少带有小钢珠的笔头的更换，如图 14–8 所示。

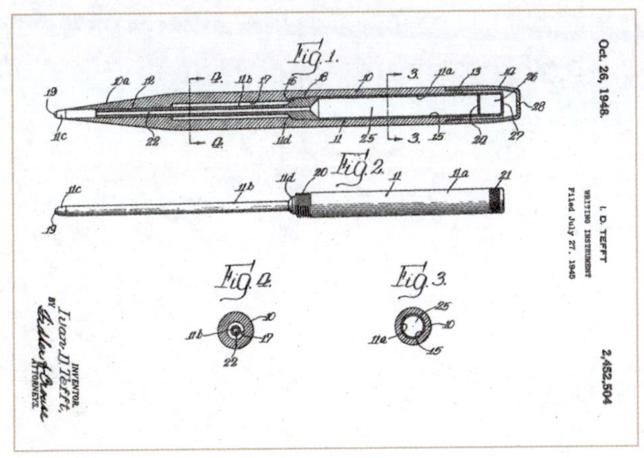

图 14–8

为了使圆珠笔芯能够在不书写的时候退回到笔杆中，避免油墨通过小钢珠的滚动漏出从而弄脏衣物，1950 年派克公司再次改进并获得美国 US2519341 号专利权，这一次的改进之处是在笔芯的末端安装了弹簧结构，如图 14-9 所示。

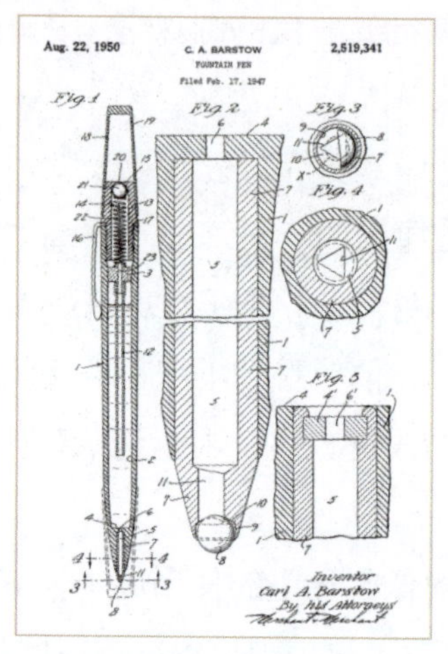

图 14-9

1954 年，派克钢笔公司推出了高级的 Jotter 圆珠笔，这种笔容量大而且能在油墨用完后补充油墨。自 1954 年以来，Jotter 圆珠笔的全球销量累计已超过 7.5 亿支，已经成为目前全球销量最大的圆珠笔。

圆珠笔虽小，但是这项发明的作用和意义却非常之大，圆珠笔的发明被誉为制笔工业的一场革命。阿根廷人为了表示对比克的尊敬，把他的生日 9 月 29 日定为“发明家节”。

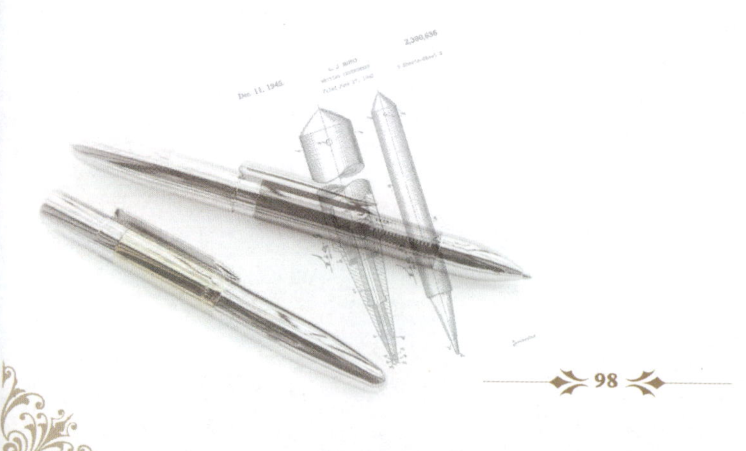

1945.US 2390636

书写工具

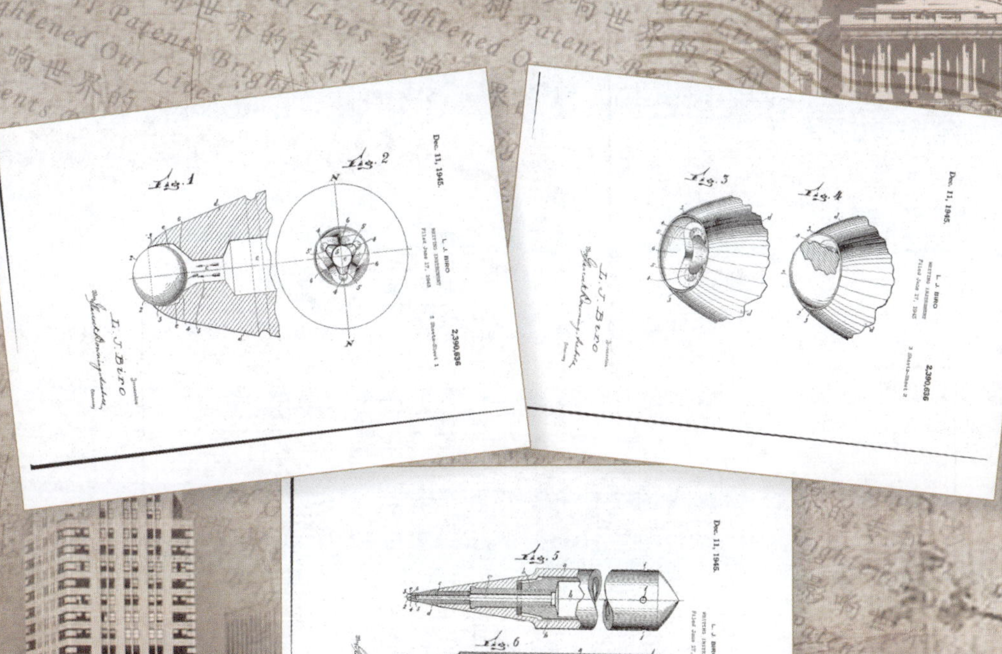

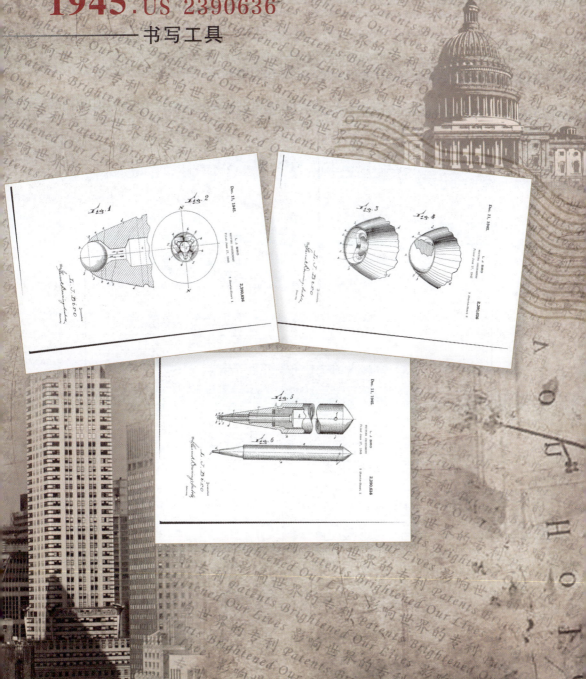

美国 America

Patented Dec. 11, 1945 2,390,636

UNITED STATES PATENT OFFICE

2,390,636

WRITING INSTRUMENT

Laszlo Jozsef Biro, Buenos Aires, Argentina.

Application June 17, 1943, Serial No. 491,206
In Argentina April 17, 1943

11 Claims. (Cl. 120—43)

This invention relates to writing instruments, and particularly to a feed system for fountain pens having a freely rotatable ball at the writing end thereof.

Fountain pens have generally given good results, but inasmuch as the pen is open and keeps the ink in permanent contact with the atmosphere, the evaporation of the ink cannot be avoided, and this makes it necessary to frequently refill the ink reservoir.

Attempts have been made to provide a fountain pen having a closed terminal, and particularly a free ball nib, which partially overcomes the above-mentioned drawbacks, but so far it has not been possible to obtain a suitable feed, regardless of the different types of ink used, such as the pasty ink combined with driving means for causing same to reach the ball at will.

The ideal construction in this type of instrument is undoubtedly the ball terminal system, with a suitable setting or socket so as to constitute a substantially air-tight closure but provided with means which, without the necessity of pressure or driving means, will maintain a charge of ink in the feed conduit, so that through suitable branches, said ink will be in constant contact with the surface of said ball within said socket which will thus remain permanently ready for use, and this is precisely what is attained by means of the present invention.

The present invention comprises a handy instrument provided with a reservoir for ink which is capable of being fed by simple gravity and capillarity, the main feature thereof being constituted by the feed conduit which is so combined that the flow of ink, instead of being stopped at the ball, extends against the surface thereof to the vicinity of the ball setting.

For this purpose, upon reaching the ball or the neighbourhood thereof, said conduit is branched by means of channels or grooves extending to the sides of the ball setting, and thus the ball is kept practically covered with ink, so that on rolling out the inner surface thereof, the ink will mark the paper with well defined strokes. The continuity of the displacement will not alter the regularity required for writing, since the branches of the channel are capable of holding a uniform and constant charge of ink.

One of the objects of the present invention is to provide means for obtaining a charge which will last longer due to the arrangement of the reservoir which enables the use of an ink which will not evaporate in mass.

A further object of the invention is to maintain a free and regular ink feed, inasmuch as due to the branches of the feed conduit, the ball is always kept in contact with the ink, regardless of the pressure exerted thereon.

A still further object of the invention is to provide an instrument which can be readily used without requiring special mechanisms, inasmuch as the ink will flow by gravity, without the aid of any auxiliary means.

Other objects and advantages of the invention will become apparent from the course of the following description, when read in conjunction with the accompanying drawings showing, by way of example, a preferred embodiment of the invention. In the drawings:

Fig. 1 is a view of the tip or point of the instrument, taken along the line N—N of Fig. 2, so as to show the capillary channel and the branches thereof adjacent to the ball.

Fig. 2 is a front view of the terminal with the ball omitted so as to clearly show the structure of the ball seat which provides a suitable bearing, and also the channels constituting the branch from the feed conduit.

Fig. 3 is a perspective view of the instrument terminal or tip, the ball having been omitted so as to show the ball seat structure providing the necessary bearing points, and also the branch channels of the feed conduit.

Fig. 4 is a perspective view of the same assembly of Fig. 3, but including the ball.

Fig. 5 is a view of the instrument with the front portion in longitudinal section in order to show the arrangement of the conduit or channel constituting the feeder for the terminal ball.

Fig. 6 is an external view of the complete instrument.

The same reference characters are used to indicate like or corresponding parts or elements throughout the different drawings.

As may be seen from the drawings, *a* is the body of the writing instrument having therein a reservoir *b* intended to contain a charge of thick or dense ink. Said reservoir *b* is provided with an air intake 1 so that the ink may descend by gravity through a feed conduit *c*.

Said feed conduit *c* leads to a tip *d* terminating in a spherical ball 2 which is intended to contact the writing paper or surface.

The cross-section of said conduit *c* is just large enough to allow the passage of ink, that is to say, that it may be given a capillary nature, so that a constant charge of liquid will be maintained by adhesion within said conduit *c*.

A concave socket *e* is provided at the end of

2 2,390,636

said conduit *c* and serves to house the ball 2 which is secured by means of a setting 3 ending in an annular line or rim extending beyond the plane corresponding to the maximum circle of said ball, but leaving free, externally of said setting, a portion of the ball sufficiently large to constitute a projecting terminal, as clearly shown in Figs. 1, 4, 5 and 6.

The concave socket *e* carries three bearing points 4, equidistant from the center thereof and constituting a seat for said ball, which is snugly held by said setting 3. As will be observed from Figs. 1 and 2, said bearing points 4 conform closely to the spherical surface of the ball 2 and extend over an appreciable area. It will also be noted that the ball is spaced from the walls of the concave socket at all points except at said bearing areas 4 and at the ball setting or rim 3.

Before reaching the ball 2, said conduit *c* is narrowed, forming a bottle-neck passage 5 which, while being cylindrical, assumes a frusto-pyramidal shape in a widened portion 5'. The channels 6 start at said widened portion adjacent the conduit *c* and are dihedral-angle grooves, the deepest lines of which are radial and occupy the areas between the bearing points 4, so that when the ball is placed in proper position, there will be a clearance in the three channels 6, as shown in Figs. 1 and 5.

Inasmuch as the channels 6 are actually an extension of the final widened portion of conduit *c*, said channels 6 constitute a branch of said conduit, and since said channels are directed towards the side cavity of concave *e*, the branches of said conduit *c* are thus extended to the vicinity of the setting 3 and therefore, with the exception of the three spherical seats determined by the bearing points 4, the ink will be in contact with the major part of the surface of said ball 2.

The device is suitable for using a very adhesive ink, capable of forming a thin layer on the surface of the ball 2, so that in rolling same, said layer may define the strokes in a very regular manner and very clearly.

The ink should fulfill other requirements, such as the ability to keep in mass without drying and without oxidizing in contact with the air.

In order to obtain a successful operation of the instrument, the cavity of the concave should be as ample as possible or at least it should be provided with grooves or sockets deriving from or combined with the end of the feeder and the branches thereof.

Operation

As mentioned hereinbefore, the operation of the instrument subject of the present invention is very simple, as the thick or dense ink charged in the reservoir *b* is derived by gravity to the conduit *c*, which will permanently contain a full charge due to the capillary or other suitable structure thereof. From said conduit *c*, the ink will pass through the channels or grooves 6 and reach the surface of the ball 2. The ink will cover a considerable portion of said ball, since apart from the seats 4, the remaining portion of the concave *e* is hollow, so as to allow the access of the ink coming through the branches or channels 6. In other words, the ink vein coming from the conduit *c*, after reaching the surface of the ball, will spread on said surface to the neighbourhood of the end or rim of the setting, and thus practically the entire portion of the ball which is housed in the setting will be kept covered with ink, so that in rolling the ball through engagement with paper or any other surface for the purpose of writing, the inked portion of the ball will come in contact with the writing surface, in such a manner that the strokes obtained are not only well defined, but also uniform. The rolling of the ball will not interrupt the ink feed, since the ink is replenished by gravity, without the necessity of auxiliary means or operations; that is to say, the discharge of the ink is automatically controlled and compensated by means of the admission of the air entering through the corresponding intake 1.

In short, the ink feed takes place by means of ink veins which are branched from the general conduit *c* and distributed through the cavities or hollow portions of the concave *e*, to the neighbourhood of the end or rim of the ball setting, so as to establish an effective contact between the ink and the ball. Due to the fact that the ink used is very adhesive, upon rolling the ball 2, sufficient ink will pass out to mark the strokes.

The seats 4 comprise three bearing points which, combined with the setting or rim 3, constitute a housing for said ball 2, the adjustment of said ball being such as to enable the free rolling thereof in any direction, while at the same time the ball will have no substantial play within its socket. It will be understood that the seats 4 may comprise four or more bearing points, instead of three as shown. Also, instead of being radial, said channels 6 may be helicoidal or of any other type, provided they derive from the general conduit *c* and lead to the concave *e*, where the ink is placed in contact with the surface of said ball 2.

In the present instance the reservoir *b* is detachable relative to the body *a*, but it will be readily understood that the device may be arranged so that the walls of the body *a* will constitute the walls of the reservoir *b*.

Also, it is evident that in carrying out the invention, several changes in construction and details will occur to those skilled in the art, without departing from the scope of the invention as clearly set forth in the appended claims.

I claim:

1. A writing instrument of the type comprising a reservoir for charging dense ink having an air-intake and a free ball tip located in a corresponding setting and fed through a single conduit derived from said reservoir, wherein said ball setting is constituted by channels provided in the walls thereof and leading from said conduit to the sides of the setting socket, said channels being spaced so as to remain in alternate relation with bearing seats for said ball in said setting.

2. A writing instrument of the type comprising a reservoir for charging dense ink having an air-intake and a free ball tip located in a corresponding setting and fed through a single conduit derived from said reservoir, wherein said ball setting is constituted by channels provided in the walls thereof and leading from said conduit to the sides of the setting socket, said channels being spaced so as to remain in alternate relation with bearing seats for said ball in said setting, the virtual axis of said single conduit concurring to the centre of said ball and being branched by means of channels formed in the seat of said free ball.

3. A writing instrument of the type comprising a reservoir for charging dense ink having an air-intake and a free ball tip located in a corresponding setting and fed through a single conduit derived from said reservoir, wherein said ball setting is constituted by radial channels provided in the walls thereof and leading from said conduit to the sides of the setting socket, said channels being spaced so as to remain in alternate relation with bearing seats for said ball in said setting.

2,390,636 3

4. A writing instrument of the type comprising a reservoir for charging dense ink having an air-intake and a free ball tip located in a corresponding setting and fed through a single conduit derived from said reservoir, wherein said ball setting is constituted by channels provided in the walls thereof and leading from said conduit to the sides of the setting socket, said channels being spaced so as to remain in alternate relation with bearing seats for said ball in said setting, said single conduit being capillary and of cylindrical section, and acquiring in the vicinity of the ball a frusto-conical shape terminating in radial channels located in alternate relation with bearing seats for said body.

5. A writing instrument as claimed in claim 4, wherein said channels are in the shape of dihedral angles, said channels ending in conjugate relation with the concave for said ball, so that the depth of said channel, relative to the surface of said ball, will form ink branch veins extending to the vicinity of the end of the setting for said ball.

6. A writing instrument of the type comprising a reservoir for charging dense ink having an air-intake and a free ball tip located in a corresponding setting and fed through a single conduit derived from said reservoir, wherein said ball setting is constituted by channels provided in the walls thereof and leading from said conduit to the sides of the setting socket, said channels being spaced so as to remain in alternate relation with bearing seats for said ball in said setting, said single conduit being provided with a bottle-neck portion or zone of smaller internal section located in the vicinity of said ball, said bottle-neck portion being branched, upon reaching said ball-bearing piece, by means of said channels.

7. A writing instrument of the type comprising a reservoir for charging dense ink having an air-intake and a free ball tip located in a corresponding setting and fed through a single conduit derived from said reservoir, wherein said ball setting is constituted by channels provided in the walls thereof and leading from said conduit to the sides of the setting socket, said channels being spaced so as to remain in alternate relation with bearing seats for said ball in said setting, said single conduit having a constant section extending to said channels.

8. A writing instrument of the type comprising a reservoir for charging dense ink and a free ball tip located in a corresponding setting and fed through a single conduit derived from said reservoir, wherein said ball setting is constituted by channels provided in the walls thereof and leading from said conduit to the sides of the setting socket, said channels being spaced so as to remain in alternate relation with bearing seats for said ball in said setting, said setting being of one-piece construction and including a ball-retaining rim in forwardly-spaced relation to said bearing seats.

9. In a writing instrument of the type having a freely-rotatable writing ball and a reservoir from which ink is fed to the ball, the improvement which comprises a socket for the ball in the front end of the instrument provided about its bottom with a plurality of circumferentially-spaced generally spherical ball seating sections with which the ball conforms over substantial areas, a single ink-feeding conduit connected with the reservoir and opening into the center of the bottom of the socket, and a plurality of ink-feeding channels in the socket between said ball seating sections, said channels opening at their rear ends into the conduit and terminating at their front ends forwardly of said sections.

10. In a writing instrument of the type having a freely-rotatable writing ball and a reservoir from which ink is fed to the ball, the improvement which comprises a socket for the ball in the front end of the instrument provided about its bottom with a plurality of circumferentially-spaced generally spherical ball seating sections with which the ball conforms over substantial areas, a single ink-feeding conduit connected with the reservoir and opening into the center of the bottom of the socket, and a plurality of ink-feeding channels in the socket between said ball seating sections, said channels opening at their rear ends into the conduit and at their front ends into the socket forwardly of said sections, said socket being provided at its edge with a ball-retaining rim in forwardly-spaced relation to said ball seating sections, and said ball being spaced from the walls of the socket at all points other than at said rim and said ball seating sections.

11. In a writing instrument of the type having a freely-rotatable writing ball and a reservoir from which ink is fed to the ball, the improvement which comprises a socket for the ball in the front end of the instrument provided about its bottom with generally spherical ball seating means with which the ball conforms over a substantial area, a single ink feeding conduit connected with the reservoir and opening into the center of the bottom of the socket, and ink-feeding channel means in the socket interrupting the continuity of said ball seating means, said channel means opening rearwardly into the conduit and terminating forwardly of said ball seating means.

LASZLO JOZSEF BIRO.

条形码的发明 Barcode

乔·伍德兰德和伯尼·西尔沃

Joe Wood Land、Bernard Silver

1949年，美国乔·伍德兰德和伯尼·西尔沃两位工程师开始研究用代码表示食品项目及相应的自动识别设备，并获得了美国专利（US2612994）。该图案很像微型射箭靶，被叫作“公牛眼”代码，在原理上，“公牛眼”代码与后来的条形码很相近，遗憾的是当时的工艺和商品经济还没有能力印制出这种码。

条形码最早出现在40年代，但是得到实际应用和发展还是在70年代左右。现在世界上的各个国家和地区都已经普遍使用条形码技术，而且它正在快速的向世界各地推广，其应用领域越来越广泛，并逐步渗透到许多技术领域。

US 2612994

www.sipo.gov.cn

第十五篇 条形码

条形码技术的研究最早可追溯至 20 世纪 20 年代，威斯汀豪斯的实验室里，一位名叫约翰·科芒德（John Kermode）的发明家“异想天开”地想对邮政单据实现自动分检。他的想法是在信封上做条码标记，条码中的信息是收信人的地址，就像今天的邮政编码。为此科芒德发明了最早的条码标识（简称科芒德码），设计方案非常简单，即一个“条”表示数字“1”，二个“条”表示数字“2”，以此类推。然后，他又发明了由基本元件组成的条码识读设备：一个扫描器，能够发射光并接收反射光；一个测定反射信号条和空的方法，即边缘定位线圈；以及使用测定结果的方法，即译码器。

科芒德的扫描器利用当时新发明的光电池来收集反射光。“空”反射回来的是强信号，“条”反射回来的是弱信号。与当今高速度的电子元器件应用不同的是，科芒德利用磁性线圈来测定“条”和“空”。就像一个小孩将电线与电池连接再绕在一颗钉子上来夹纸。科芒德用一个带铁芯的线圈在接收到“空”的信号的时候吸引一个开关，在接收到“条”的信号的时候，释放开关并接通电路。因此，最早的条码阅读器噪声很大。开关由一系列的继电器控制，“开”和“关”由打印在信封上“条”的数量决定。通过这种方法，条码符号直接对信件进行分检。

然而，科芒德码所包含的信息量相当的低，并且很难编出 10 个

以上的不同代码。此后不久，科芒德的合作者道格拉斯·杨（Douglas Young）在科芒德码的基础上作了些改进。杨码使用更少的条，但是利用条之间空的尺寸变化，可在同样大小的空间对 100 个不同的地区进行编码。

条形码的相关专利最早出现在20世纪40年代，当时，来自美国的两位工程师乔·伍德兰德（Joe Wood Land）和伯尼·西尔沃（Bernard Silver）着手研究用代码表示食品项目的方法及相应的自动识别设备，并于1949年申请了名为“分级装置和方法”的美国US2612994号专利，如图15-1、图15-2所示。

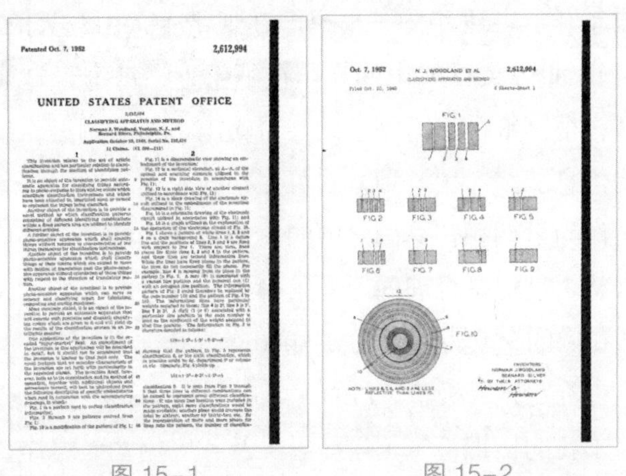
Patented Oct. 7, 1952　　2,612,994

UNITED STATES PATENT OFFICE

图 15-1　　图 15-2

该专利的图案很像微型射箭靶，被称为“公牛眼”代码。靶的同心圆是由“条”和“空”绘成的圆环形。发明人的想法是利用科芒德码和杨码的垂直的“条”和“空”，并使之弯曲成环状，非常像射箭的靶子。这样扫描器通过扫描图形的中心，能够对条形码符号解码，不管条形码符号方向的朝向。在原理上，“公牛眼”代码与后来的条形码很相近。

十年后，乔·伍德兰德作为 IBM 公司的工程师成为北美统一代码 UPC 码的奠基人。以吉拉德·费伊塞尔（Girard Fessel）为代表的几名发明家于 1959 年提请了一件专利，描述了数字 0 ～ 9 中每个数字可由 7 段平行条组成。但是这种码机器难以识读，人读起来也不方便。

不过这一构想促进了后来条形码的产生与发展。不久，E.F. 布宁克（E.F.Brinker）申请了另一件专利，该专利是将条码标识在有轨电车上。20 世纪 60 年代期西尔沃尼亚（Sylvania）发明的一个系统，被北美铁路系统采纳。这两项可以说是条形码技术最早期的应用。此后，截至 2014 年，受“公牛眼”专利的启发，出现了近百项有关条形码的生产或读取设备的发明。随着条形码生产与读取技术的发展，条形码的应用也逐渐开始发展起来。

1970 年，易腾迈公司开发出了“二维码”，并有了价格适于销售的二维矩阵条码的打印和识读设备。那时二维矩阵条形码用于报社排版过程的自动化。二维矩阵条形码印在纸带上，由今天的一维 CCD 扫描器扫描识读。CCD 发出的光照在纸带上，每个光电池对准纸带的不同区域，根据纸带上印刷条码与否输出不同的图案，组合产生一个高密度的信息图案。用这种方法可在相同大小的空间打印上一个单一的字符，作为早期科芒德码之中的一个单一的条。定时信息也包括在内，所以整个过程是合理的。当第一个系统进入市场后，包括打印和识读设备在内的全套设备大约要 5000 美元。

此后，随着发光二极管（LED）、微处理器和激光二极管的不断发展，迎来了新的标识符号（象征学）及其应用的大爆炸，人们称之为“条码工业革命”。

时至今日，在网络世界里，每个人、每件商品都有属于自己的条形码，而手机摄像头就能扫码，这足以证明条形码出现的深远意义。

美国 America

1949.US 2612994

——分级装置和方法

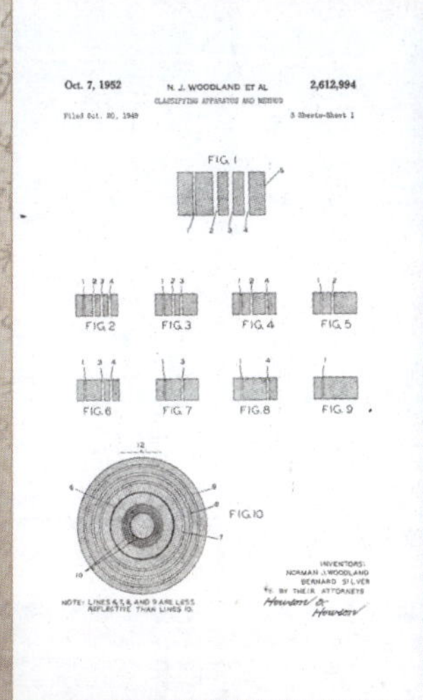

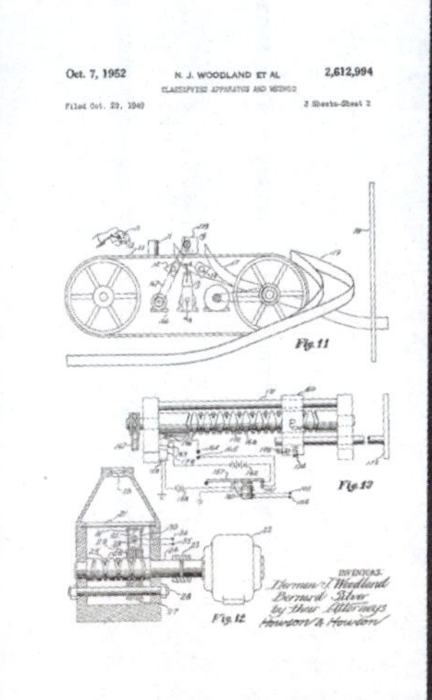

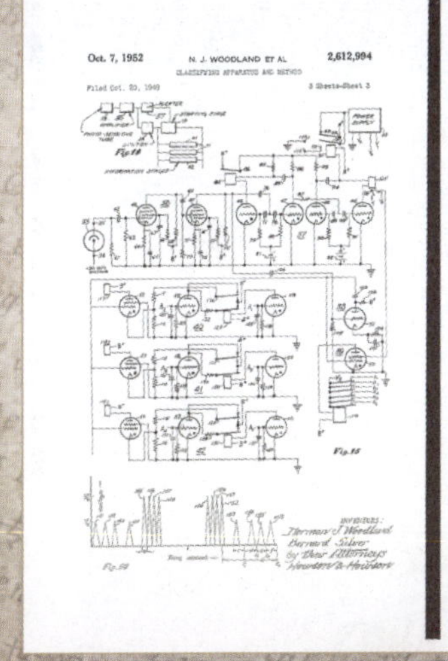

Patented Oct. 7, 1952

2,612,994

UNITED STATES PATENT OFFICE

汽车安全带的发明

LAP-BELT

尼尔斯·博林 Nils Bohlin

尼尔斯·博林于1959年发明了三点式安全带，享誉全球。今天，他的发明已作为标准配置安装在全球所有的汽车内，并在世界各地的交通事故中拯救了数百万生命。

当时的对角线式安全带在高速撞击下达不到安全标准，而博林设计的汽车安全带能同时跨过腹部以下部位并横跨在肩部，即横跨在骨盆和胸腔之上，通过位于座椅一侧的低位固定点协调发挥作用。1961年“V型三点式安全带”获得德国专利DE1101987。同年沃尔沃汽车公司在世界上率先采用三点式安全带，首次将这项专利产品引入北欧市场。

撞击试验结果表明，三点式安全带给予乘员的保护作用最为杰出。事实证明，这个貌似简单的发明开创了汽车安全的新时代。在人们心里ABS、气囊是汽车安全性能的代言词，但在发生交通事故时，对人们的安全起最主要保障作用的却是汽车安全带。

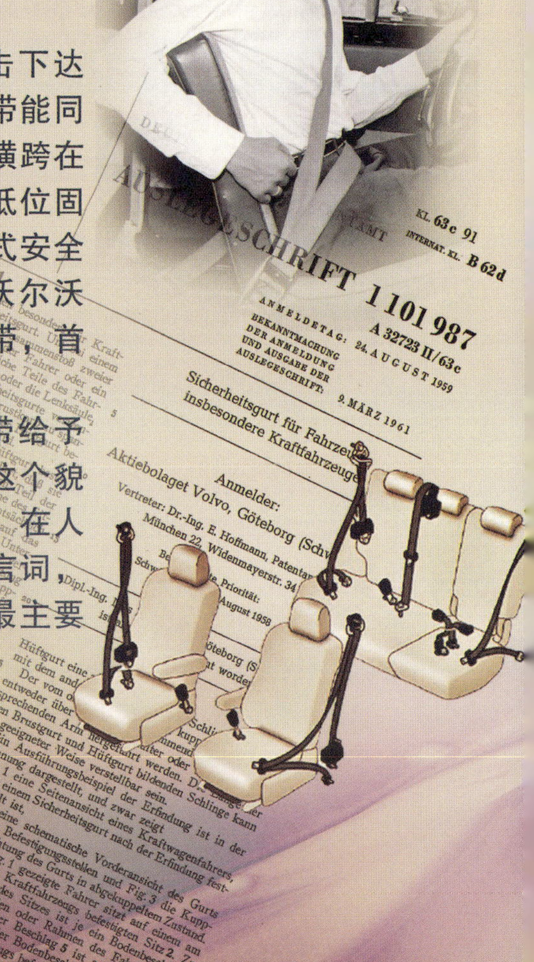

DE 1101987

www.sipo.gov.cn

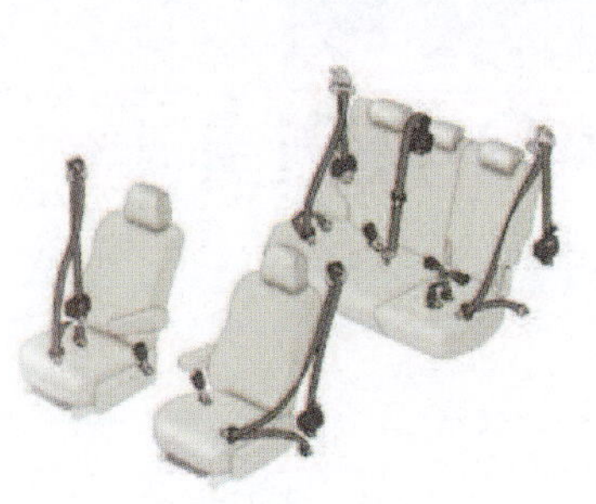

第十六篇 汽车安全带

19 世纪 80 年代末，伴随着第一次工业革命的热潮，汽车出现在人们的视野中，是人类第一次摆脱人力、畜力，利用动力驱动交通工具。此后的二三十年中，汽车工业的发展在欧洲以及美国呈现出一个高潮，尤其在动力驱动、速度提升、车体结构等方面都有显著的突破并迅速发展。汽车安全带就是这其中的突破之一。

20 世纪初叶，已经有人意识到汽车驾驶中的安全隐患，并着手进行安全带的研究，同时在“二战”期间，美军也为克服载人军用飞机飞行中乘员的固定问题在飞机上安装安全带。此外，在 20 世纪 50 年代，一位来自美国加利福尼亚的神经专科医生，因为发现很多前来就诊的患者都是因为车祸而造成脑部损伤，于是便研发了可伸缩的安全带，并在 1955 年 11 月 5 日发表的《美国医药协会日志》（JAMA）上公开了此项研究。然而这些发明并没有受到广泛认可与应用。

直到 1959 年，瑞典人尼尔斯·博林（Nils Bohlin，1920–2002 年）发明了“三点式”安全带并先后获得德国专利权和美国专利权。从专利文献中我们发现，博林首先在本国提出专利申请，而后以 1958 年瑞典优先申请分别向德国和美国提交了专利申请，并于 1961 年获得名为“车用安全带，尤其是机动车辆”的德国 DE1101987 号专利权，如图 16–1、图 16–2 所示，1962 年获得美国 US3043625 号专利权。

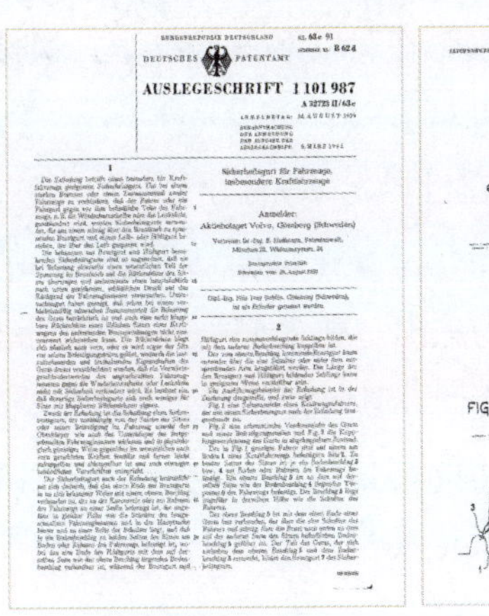

DEUTSCHES PATENTAMT

AUSLEGESCHRIFT 1 101 987

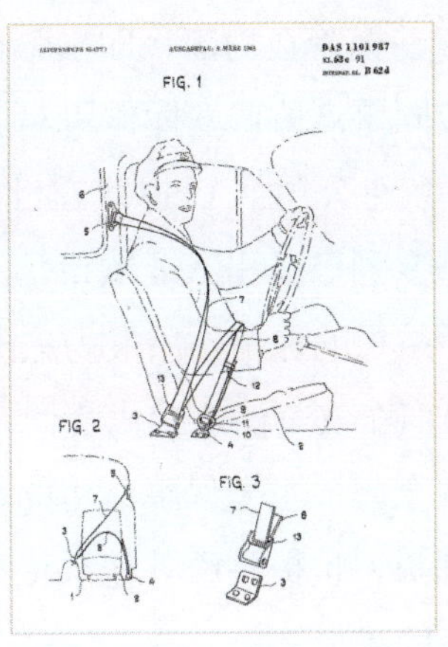

图 16-1　　图 16-2

1958 年，博林受雇于沃尔沃汽车公司，成为该公司的首席安全工程师，负责新型汽车被动安全设施的开发。此前，博林就职于瑞典 Svenska Aeroplan AB 飞机公司（简称 SAAB 公司），研发飞机驾驶员弹射座椅并获得瑞士 CH3752369 号专利权，如图 16-3、图 16-4 所示。

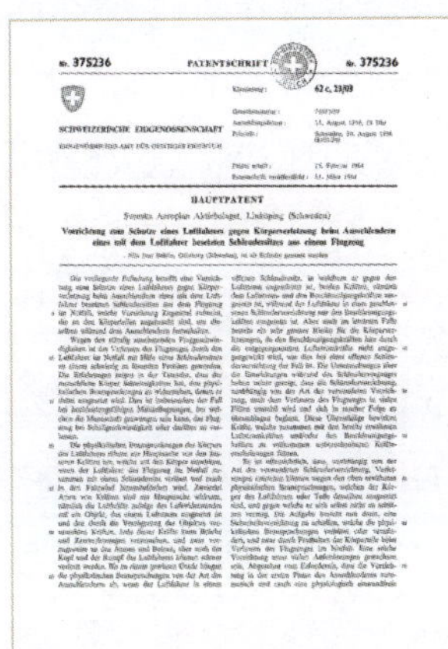

PATENTSCHRIFT

Nr. 375236

SCHWEIZERISCHE EIDGENOSSENSCHAFT

HAUPTPATENT

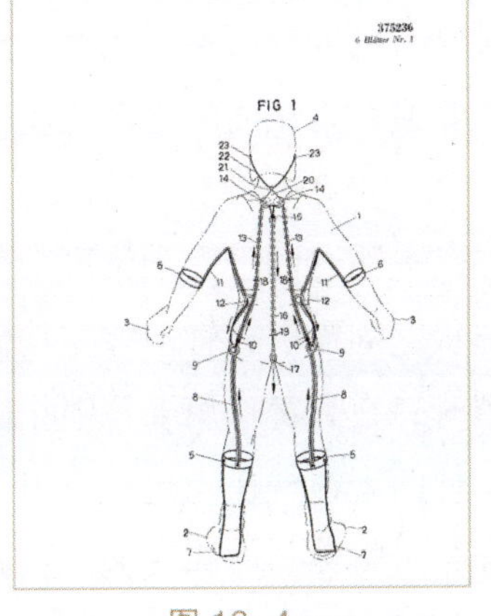

图 16-3　　图 16-4

在此基础上，博林对公路事故及汽车安全方面也有其独到的见解，并掌握了处理这类问题的专业技能。在那个年代，普遍使用的汽车安全带是固定在座椅后面，交叉地绑在人体上，并在腹部用搭扣锁定。但是，这种“对角线式”的安全带在高速撞击下无法阻止人体活动，达不到安全标准。而博林提出的“三点式”安全带，是能同时跨过腹

部以下部位并横跨在肩部，即横跨在骨盆和胸腔之上，通过位于座椅一侧的低位固定点协调发挥作用。

沃尔沃汽车公司于1959年在世界上率先采用“三点式”安全带，首次将这项专利产品引入北欧市场。撞击试验结果表明，“三点式”安全带给予乘员的保护作用最为杰出。在人们心里ABS制动防抱死系统、气囊是汽车安全性能的代名词，但研究表明，在发生交通事故时，对人们的安全起最主要保障作用的却是汽车安全带。这种安全带不仅提供给驾驶员，也提供给其他乘位人员。事实上就是现在广泛使用的汽车安全带的雏形。

通过专利说明书附图，如图16-2所示，可以清晰地看到，此项发明设计是在驾驶员座位两侧分别设有与车辆底板或框架牢固相连的安装装置，装置的座位同侧的车辆门柱上设置了一个顶部安装装置。该顶部安装装置设在驾驶员肩水平面上方。安全带一端与顶部安装装置相连接，安全带的另一端经过驾驶员一个肩并斜跨其胸部，穿过与所在安装装置上的锁扣件上的槽，形成安全带的部分；然后穿过借助于枢轴与安装装置相连的可调节皮带长度的环扣，形成横跨髋部的安全带的部分；扣环上设有可移动的滚花锁栓，以便皮带调节到相应长度后锁住，滑圈有助于安全带剩余部分与安全带的部分紧密相贴。锁扣件两边板之间设有枢轴，并在其上设有碰锁控制杆，当锁扣件插入安装装置时，在弹簧作用下，锁扣件的两凸起物锁紧在安装装置的两孔眼中，使两者锁合在一起；当用手操作碰锁控制杆就可使凸起物与孔眼脱开，可将锁扣件和安装装置分离。博林的发明巧妙的运用了力学原理，用三个受力点建立了一个受力面，从而提升安全带的保护性能。此后，关于汽车安全带的创新依然在继续，而博林发明的三点安全带为此奠定了坚实的基础。尤其在1964～1985年的20年间，多项相关汽车安全带的专利发明在创新中都是基于博林这项发明的技术内容进行的再创新。

现如今，在人们的日常生活中，汽车上普遍采用的都是三点安全

带。同时，随着家用汽车的不断普及，儿童安全座椅的使用比例也在逐步提高，有些国家甚至将儿童安全座椅的使用写入法律。儿童安全座椅使用的是五点安全带，相较三点安全带，五点安全带在人体两个肩膀及大腿两侧都设计了保护受力点，能够更好的保证受力与保护的平衡性。五点安全带也被应用于赛车中。此外，还有一些具有特殊用途的多点安全带，但都没有脱离三点安全带的最基本功能。

图 16-5 儿童安全座椅

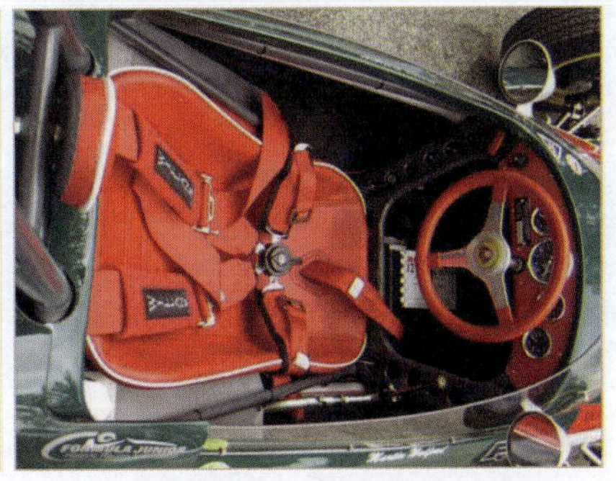

图 16-6 赛车座椅

尼尔斯的发明于 1985 年被德国专利局评为过去 100 年最重要的八大发明之一，同时也入选了由美国纽约大学出版社和巴恩斯诺布尔出版社于 2007 年联合出版的《20 世纪影响世界的 100 个发明》之中。

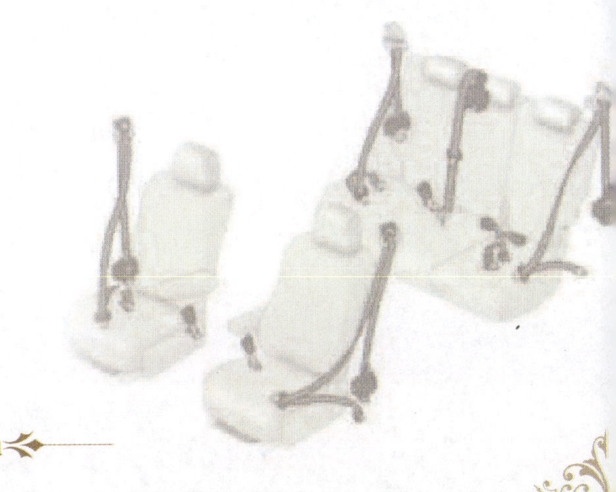

1961.DE 1101987

——车用安全带，尤其是机动车辆

BUNDESREPUBLIK DEUTSCHLAND

DEUTSCHES  PATENTAMT

KL. 63c 91

INTERNAT. KL. B 62d

AUSLEGESCHRIFT 1 101 987

A 32723 II/63c

ANMELDETAG: 24. AUGUST 1959

BEKANNTMACHUNG DER ANMELDUNG UND AUSGABE DER AUSLEGESCHRIFT: 9. MÄRZ 1961

Sicherheitsgurt für Fahrzeuge, insbesondere Kraftfahrzeuge

Anmelder:
Aktiebolaget Volvo, Göteborg (Schweden)

Dipl.-Ing. Nils Ivar Bohlin, Göteborg (Schweden), ist als Erfinder genannt worden

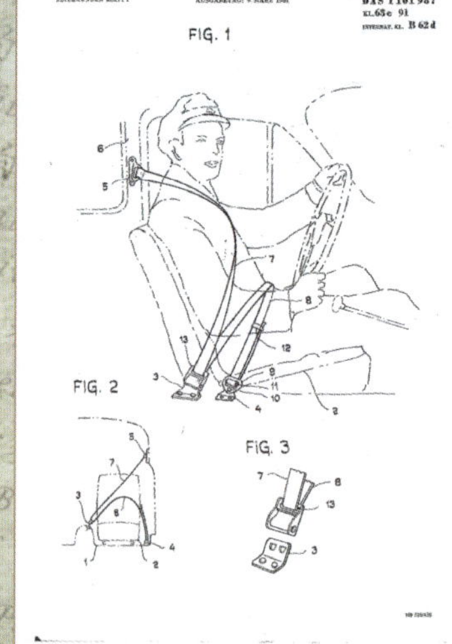

鼠标的发明
Mouse invention
道格拉斯·恩格尔巴特
Douglas Carl Engelbart
鼠标是1964年由道格拉斯·恩格尔巴特（Douglas Carl Engelbart）发明的。20世纪60年代初，恩格尔巴特在参加一个会议时，随手画出了一种在底部使用两个互相垂直的轮子来跟踪动作的装置草图，这就是鼠标的雏形。1964年，恩格尔巴特再次对这种装置的构思进行完善，并动手制作出了第一个成品。
由于该装置像老鼠一样拖着一条长长的连线，像老鼠的尾巴。因此恩格尔巴特和他的同事在实验室里把它戏称为“Mouse”。
恩格尔巴特当时也曾想到将来鼠标有可能会被广泛应用，所以在申请专利时将其命名为“显示系统X-Y位置指示器”，只是人们觉得“Mouse”这个名字更为亲切，于是就有了“鼠标”这个称呼。因此，道格拉斯·恩格尔巴特也被称为“鼠标之父”。
3,541,541
Patented Nov. 17, 1970
US 3541541
www.sipo.gov.cn

第十七篇 鼠 标

20世纪60年代，计算机技术的发展还处于较为原始的阶段，根本没有现今的视窗操作系统，一切操作都是通过键盘输入复杂的指令，其烦琐程度可想而知。因此，在1967年已陆续有专利公开了纵横坐标定位技术，核心是将实际的机械运动转换为脉冲信号串，并存储在磁盘或磁带上，之后再向计算机输入，从而实现控制光标的目的。然而，虽然定位技术以及存储和转换技术已经出现，但操作及转换过程依然烦琐，不能满足实际操作的需要。

美国发明家道格拉斯·恩格尔巴特（Douglas Carl Engelbart，1925–2013年），多年来致力于开发基于计算机的技术，以提高人类对信息直接进行处理的能力。当时，他与研发团队在计算机技术研究过程中，只能靠唯一的输入设备“键盘”与计算机间进行指令的交互，哪怕是简单的移动一下光标，都必须输入键盘指令。

为突破这个局限，恩格尔巴特及其技术团队在1963年设计出了鼠标的最初原型，于1967年向美国提交名为“显示系统X–Y坐标位置指示器”的专利申请并于1970年获得美国US3541541号专利权，如图17–1、图17–2所示。该发明的原理是利用鼠标移动时带动枢轴转动，并带动变阻器改变阻值来产生位移信号，信号经计算机处理，屏幕上的光标就可以移动。这对当时来说是个神奇的突破，通过移动鼠标实现对计算机的控制简单方便，代替了烦琐的代码指令，简化了操

作的同时也为后来计算机输入、控制技术发展奠定了基础。

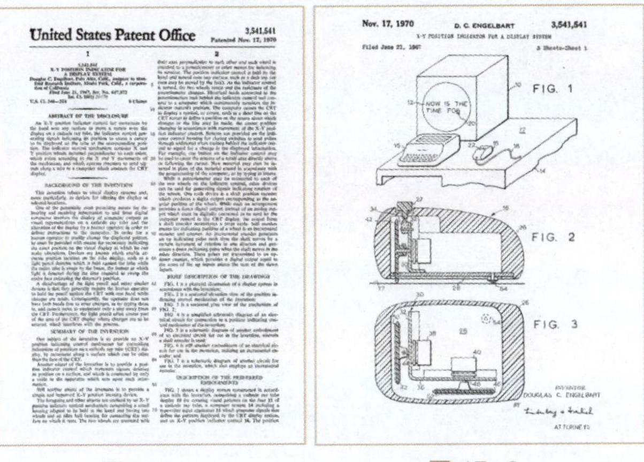
United States Patent Office 3,541,541
Patented Nov. 17, 1970

图 17–1　　图 17–2

回顾鼠标技术的发展，恩格尔巴特的鼠标其实只是作为技术验证品而存在，并未真正量产。在后续的发展中，鼠标伴随着计算机技术的发展而不断突破，进行了几次技术革新，陆续出现了机械鼠标、光电机械式鼠标及光电式鼠标。使用者握在手中的鼠标，操控性越来越好，体积越来越小。在原始鼠标专利的启发下，出现了诸如无线鼠标、带有扫描功能的鼠标等多用途产品。一时间，随着电脑在全世界占领了人们的书桌，成为每天学习工作的必备工具，鼠标也被千千万万的人握在手中。

机械鼠标　　早期鼠标　　光学鼠标　　激光鼠标

图 17–3

相比鼠标这个具体产品的发展而言，恩格尔巴特提出的“通过纵横方向定位光标技术”更具有深远的历史意义。它给予后来的研究者们相当大的启发。1972 年，美国发明人海格（HAIG）将恩格尔巴特的鼠标专利构思应用于名为“一种交互系统的信号输入控制器”的美

国 US3798599 号专利中。此后，诸如施乐、理光、苹果等多家商业公司均从鼠标的基础技术原理中得到启发，将恩格尔巴特的鼠标专利构思应用于多个技术领域。

20世纪70年代，施乐公司致力于Alto计算机系统的开发，其中相关光标控制的研究受恩格尔巴特鼠标专利技术的启发很大。1973年施乐公司申请了美国US3835464号专利，如图17-4、图17-5所示，该专利公开了一种显示系统的位置指示器，该指示器在可视显示设备表面控制光标。

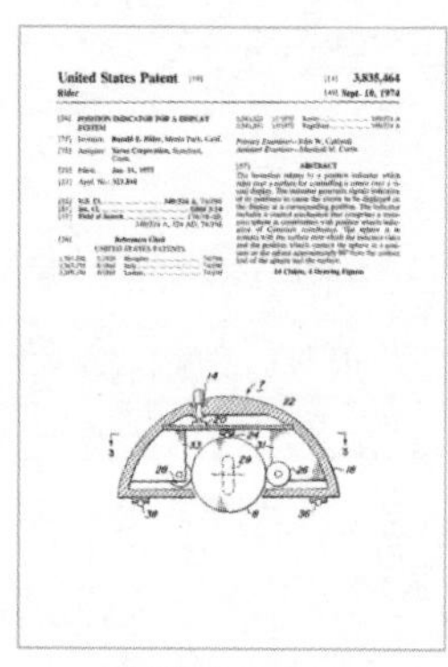

United States Patent [19] [11] 3,835,464
Rider [45] Sept. 10, 1974

图 17-4

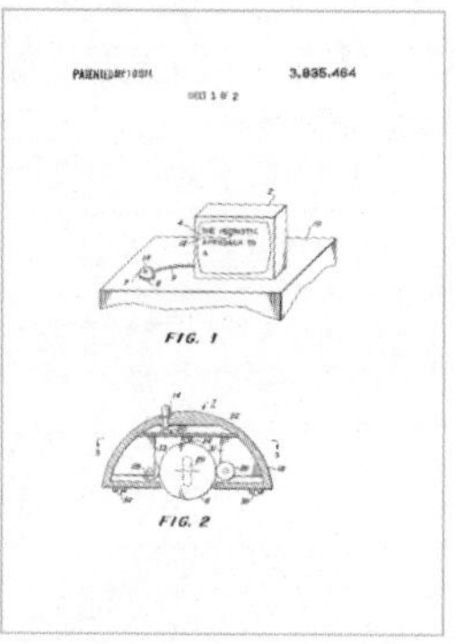

图 17-5

苹果公司一直注重通过专利制度保护自己的研发成果，并拥有大量专利。1982 年苹果公司申请了名为“一种用于显示系统的光标控制设备”的美国 US4464652 号专利，如图 17-6、图 17-7 所示，从专利

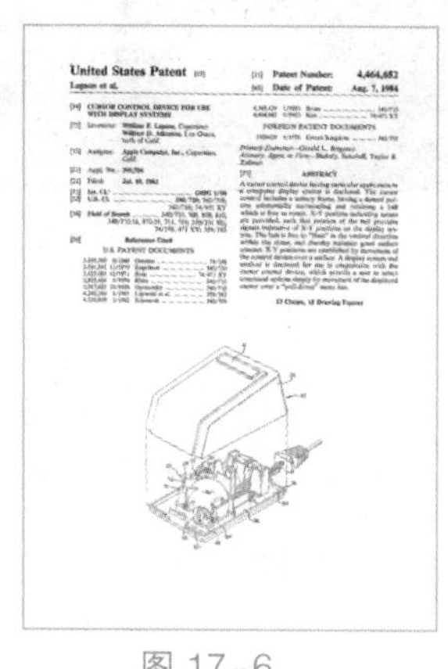

United States Patent [19] [11] Patent Number: 4,464,652
Lapson et al. [45] Date of Patent: Aug. 7, 1984

图 17-6

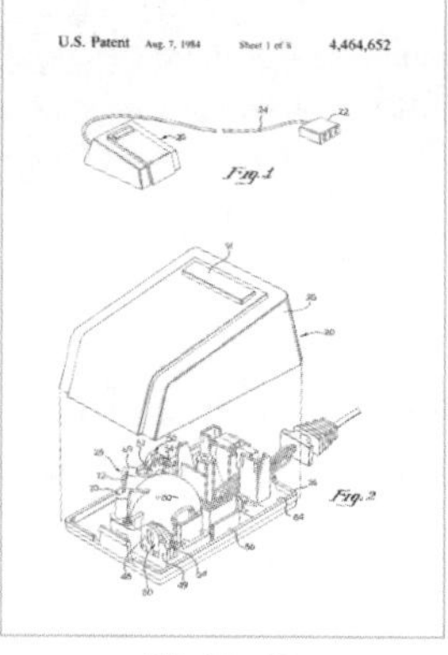

图 17-7

说明书及附图中可以看出，该项专利参考了恩格尔巴特的发明，并在随后的 30 多项专利申请中都参考了原始鼠标的技术内容。

2004 年，苹果公司申请的美国 US7663607 号专利中首次公开了多点触摸控制技术，如图 17-8、图 17-9 所示，在该专利中苹果公司再次参考了恩格尔巴特的鼠标技术内容，而这正是第一代 iPhone 研究的核心部分。

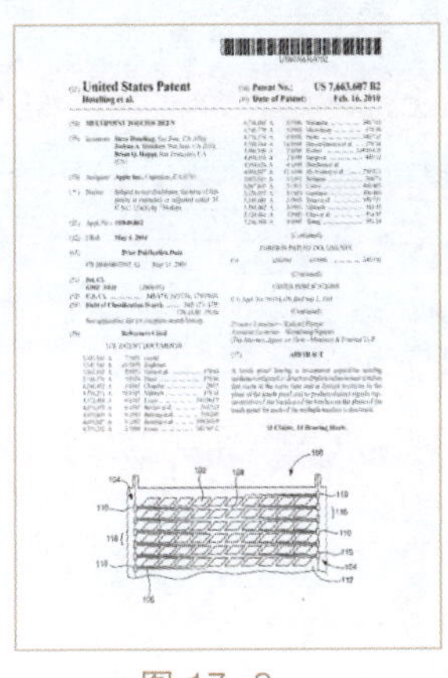

United States Patent
US 7,663,607 B2
Feb. 16, 2010

图 17-8

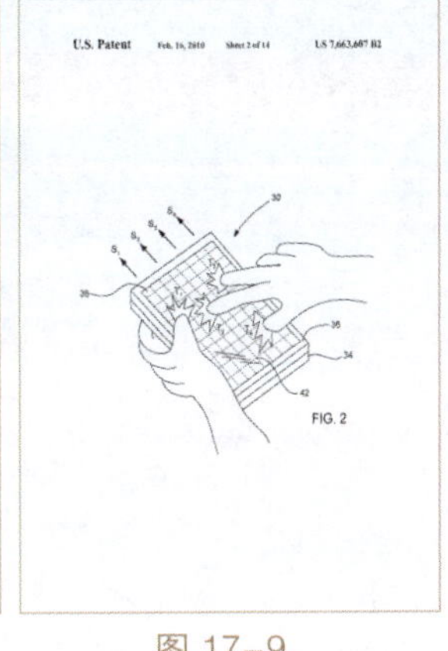

U.S. Patent Feb. 16, 2010 Sheet 2 of 14 US 7,663,607 B2

图 17-9

苹果公司在后续的十多年中不断深入探索，申请了数量可观的专利，将触控技术推进至今。似乎在一夜之间，我们已经习惯了用手指直接操控电脑、手机。与曾经独领风骚的鼠标相比，如今的触摸控制技术依然在不断进步，最大的突破就在于实现了多点控制。通俗地说，无论鼠标本身通过怎样的技术实现移动，它始终只能控制屏幕里箭头对准的那一个点。而触摸控制技术则完全不同，用两个手指即可灵活缩放屏幕上的图片就是最简单的例子。

科技发展，最根本的推动力就是想象力。

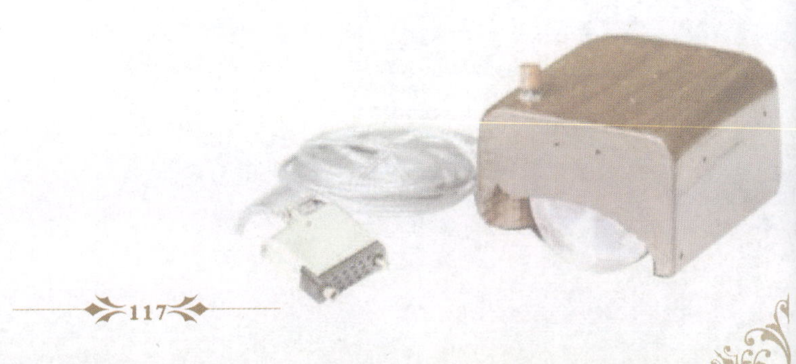

1970.US 3541541

——显示系统 X-Y 坐标位置指示器

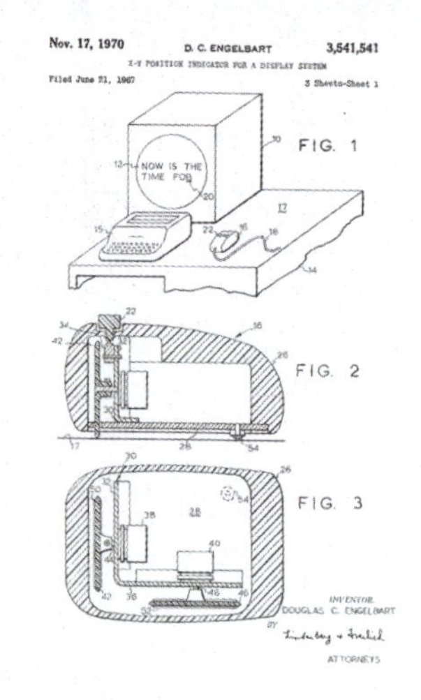

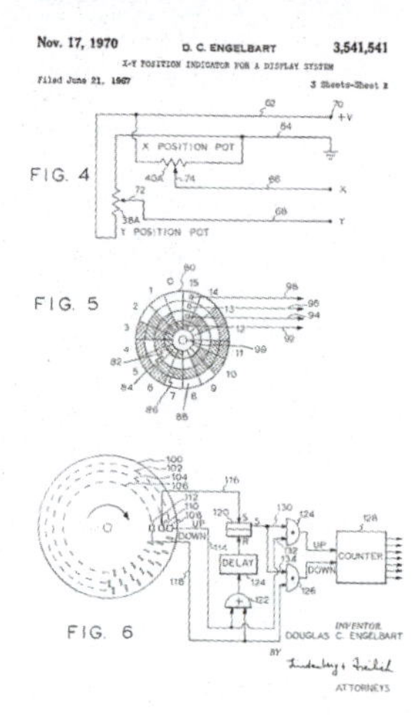

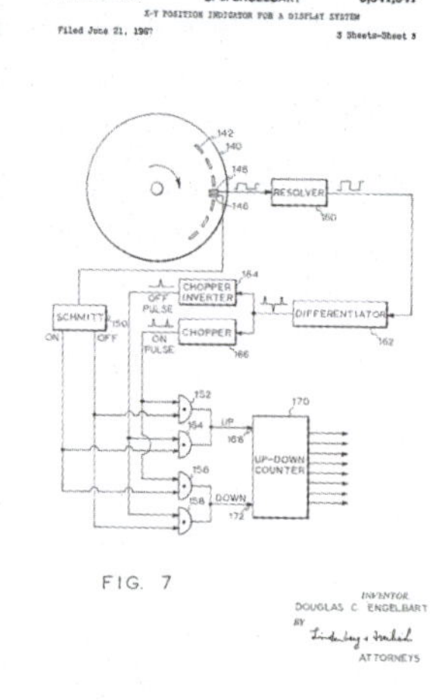

美国

America

United States Patent Office

3,541,541

Patented Nov. 17, 1970

1

3,541,541

X-Y POSITION INDICATOR FOR A DISPLAY SYSTEM

Douglas C. Engelbart, Palo Alto, Calif., assignor to Stanford Research Institute, Menlo Park, Calif., a corporation of California

Filed June 21, 1967, Ser. No. 647,872

Int. Cl. H01j 29/70

U.S. Cl. 340—324 8 Claims

ABSTRACT OF THE DISCLOSURE

An X-Y position indicator control for movement by the hand over any surface to move a cursor over the display on a cathode ray tube, the indicator control generating signals indicating its position to cause a cursor to be displayed on the tube at the corresponding position. The indicator control mechanism contains X and Y position wheels mounted perpendicular to each other, which rotate according to the X and Y movements of the mechanism, and which operate rheostats to send signals along a wire to a computer which controls the CRT display.

BACKGROUND OF THE INVENTION

This invention relates to visual display systems and, more particularly, to devices for altering the display at selected locations.

One of the potentially most promising means for delivering and receiving information to and from digital computers involves the display of computer outputs as visual representations on a cathode ray tube and the alteration of the display by a human operator in order to deliver instructions to the computer. In order for a human operator to readily change the displayed pattern, he must be provided with means for accurately indicating the exact position on the visual display at which he can make alterations. Devices are known which enable accurate position location on the tube display, such as a light pencil detector which is held against the tube while the entire tube is swept by the beam, the instant at which light is detected during the time required to sweep the entire face indicating the detector's position.

A disadvantage of the light pencil and other similar devices is that they generally require the human operator to hold the pencil against the CRT with one hand while changes are made. Consequently, the operator does not have both hands free to enter changes, as by typing them in, and cannot move to equipment only a step away from the CRT. Furthermore, the light pencil often covers part of the area of the CRT display where changes are to be entered, which interferes with the process.

SUMMARY OF THE INVENTION

One subject of the invention is to provide an X-Y position indicating control mechanism for controlling indications of positions on a cathode ray tube (CRT) display, by movement along a surface which can be other than the face of the CRT.

Another object of the invention is to provide a position indicator control which transmits signals defining its position on a surface, and which is connected by only a cable to the apparatus which acts upon such information.

Still another object of the invention is to provide a simple and improved X-Y position locating device.

The foregoing and other objects are realized by an X-Y position indicator control mechanism comprising a small housing adapted to be held in the hand and having two wheels and an idler ball bearing for contacting the surface on which it rests. The two wheels are mounted with

2

their axes perpendicular to each other and each wheel is attached to a potentiometer or other means for indicating its rotation. The position indicator control is held by the hand and moved over any surface, such as a desk top (or even may be moved by the feet). As the indicator control is moved, the two wheels rotate and the resistance of the potentiometer changes. Electrical leads connected to the potentiometers trail behind the indicator control and connect to a computer which continuously monitors the indicator control's position. The computer causes the CRT to display a symbol, or cursor, such as a short line on the CRT screen to define a position on the screen about which changes or the like may be made, the cursor position changing in accordance with movement of the X-Y position indicator control. Buttons are provided on the indicator control housing for closing switches to send pulses through additional wires trailing behind the indicator control to signal for a change in the displayed information. For example, one button on the indicator control may be used to cause the erasure of a small area directly above or following the cursor. New material may then be inserted in place of the material erased in accordance with the programming of the computer, as by typing in letters.

While a potentiometer may be connected to each of the two wheels on the indicator control, other devices can be used for generating signals indicating rotation of the wheels. One such device is a shaft position encoder which produces a digita output corresponding to the angular position of the wheel. While such an arrangement provides a direct digital output, instead of an analog output which must be digitally converted to be used by the computer control in the CRT display, the output from a shaft encoder necessitates a large cable. Still another means for indicating position of a wheel is an incremental encoder and counter. An incremental encoder generates an up indicating pulse each time the shaft moves by a certain increment of rotation in one directon and generates a down indicating pulse when the shaft moves in the other direction. These pulses are transmitted to an up-down counter, which provides a digital output equal to the sums of the up inputs minus the sum of the down inputs.

BRIEF DESCRIPTION OF THE DRAWINGS

FIG. 1 is a pictorial illustration of a display system in accordance with the invention;

FIG. 2 is a sectional elevation view of the position indicating control mechanism of the invention;

FIG. 3 is a sectional plan view of the mechanism of FIG. 2;

FIG. 4 is a simplified schematic diagram of an electrical circuit for connection to a position indicating control mechanism of the invention;

FIG. 5 is a schematic diagram of another embodiment of an electrical circuit for use in the invention, wherein a shaft encoder is used;

FIG. 6 is still another embodiment of an electrical circuit for use in the invention, utilizing an incremental encoder; and

FIG. 7 is a schematic diagram of another circuit for use in the invention, which also employs an incremental encoder.

DESCRIPTION OF THE PREFERRED EMBODIMENTS

FIG. 1 shows a display system constructed in accordance with the invention, comprising a cathode ray tube display 10 for creating visual patterns on the face 12 of a cathode ray tube, a computer system 14 including a typewriter input apparatus 15 which generates signals that define the patterns displayed by the CRT display system, and an X-Y position indicator control 16. The position

3,541,541

3

indicator control 16 is positioned on the top of the cabinet 17 of the computer, although it can be positioned on any other surface. The indicator control 16 has wheels which support it on the cabinet and which register changes in the position thereon. A wire 18 connects the position indicator control to the computer 14 for transmitting signals indicating the position of the indicator control. The computer 14, which controls the pattern on the CRT face 12, generates signals causing the display of a line or other cursor 20 on the CRT. The position of the cursor 20 is governed by the position of the indicator control 16 determined by the computer 14 in accordance with the signals it receives from the indicator control over the wire 18.

Three buttons 22 are located on the indicator control 16 for operating switches within the indicator control to allow currents to flow through conductors of the wire 18. The switches may be used to cause changes in particular areas of the display, or for other purposes. For example, one of the buttons may be used to control the delivery of signals which command the computer 14 to operate on the portion of the pattern displayed immediately above the cursor 20, such as a single character, the particular operation being designated by inputs to the typewriter apparatus 15. Another button may be used to command the operations to be performed on the entire line of characters immediately above and to the right of the cursor 20. An operation such as "erase" may be designated by pressing a particular key on the typewriter, to cause the computer to stop the display of characters at those areas. New characters can be inserted into the display by leaving the position indicator control 16 stationary so the cursor does not move and then typing in the new characters on the typewriter 15.

The position indicator control 16 is shown more clearly in the sectional side view of FIG. 2 and the sectional plan view of FIG. 3. A housing 26 has a bottom wall 28 on which is attached a right angle bracket 30. One arm 32 of the bracket holds three pushbutton switches 34 which close circuits that cause changes in the cathode ray tube display. The pushbuttons 22 are slideably mounted in the housing 26, for movement against the switches 34 to close them.

Each arm 32 and 36 of the bracket 30 holds a potentiometer, the arm 32 holding an X position potentiometer 38 and the arm 36 holding a Y position potentiometer 40. An X position wheel 42 is fixed to a shaft 44 of the potentiometer 38, while a Y position wheel 46 is fixed to a shaft 48 of the Y position potentiometer 40. Each of the position wheels 42 and 46 project through slots 50 and 52, respectively, formed in the bottom wall 28. A ball bearing support 54 fixed to the underside of the bottom wall 28 serves as a third point of support, in addition to the two wheels 42 and 46, to stably support the indicator control on the cabinet 17 or other surface.

When the position indicator control is moved over the cabinet 17, or any other surface, the X and Y position wheels rotate. Inasmuch as the X and Y position wheels 42 and 46 are mounted on axes that are perpendicular to each other, the X position wheel 42 rotates by an amount equal to the movement in one direction which may be defined as the X direction, while the Y position wheel 46 rotates an amount equal to the movement in a perpendicular or Y direction. As the wheels move, the shafts of their respective potentiometers rotate, and the resistance of the potentiometers enable continuous measurement of the resistance, and therefore of the X and Y positions of the indicator control 16. It may be noted that in most cases multiturn potentiometers are used to enable monitoring of large movements of the indicator control, or conversely, to enable fine control.

The position indicator control may be utilized by first placing it on the cabinet 17 and moving it up or down and back or forth to cause corresponding movements in the apparent position of the cursor 20, until the cursor

4

lies in a desired position. The indicator control remains stationary so long as it is left in place; therefore the cursor 20 remains fixed without any effort of the human operator. If it is desired to move the cursor 20, the position indicator control 16 is moved in directions corresponding to the desired movements of the cursor. The resistances of the rheostats, sensed through the conductors contained in the wire 18, continually monitor the position of the indicator control and cause movement of the line cursor 20 accordingly.

FIG. 4 is a simplified schematic diagram of the electrical circuit by which the position of the indicator control 16 is monitored. Electrical conductors 62, 64, 66 and 68 represent separate leads contained in the wire 18 connecting the indicator control to the computer. A voltage +V is connected at terminal 70 for sending currents through the two rheostats or potentiometers whose resistances are indicated at 38A and 40A. One side of each potentiometer is connected to lead 64, which is grounded. The wipers 72 and 74 of the potentiometers are connected to leads 68 and 66, respectively, which in turn are connected to terminals Y and X. By noting the voltage at X and Y, relative to ground potential, the resistances of the two potentiometers and therefore the X and Y positions of the indicator control are known.

The indications of X and Y position given by the voltages at terminals X and Y are presented in analog form. A digital computer requires digital inputs and therefore, an analog-to-digital converter must be used between the X and Y terminals and the computer inputs. Two types of digital output devices for use with the indicator control are shown in FIGS. 5, 6 and 7.

FIG. 5 shows a position indication control circuit which provides a digital output. An encoding disc 80 is shown which is used to indicate the X position. The disc 80, which is a simplified illustration of the type of disc which is used in practice, is divided into four rings 82, 84, 86 and 88. The disc 80 is also divided into sixteen sectors, each indicated by a number 0 through 15. Four electrical contacts connected to wires 92, 94, 96 and 98, provide readouts. Each of the sixteen sectors of each of the four rings of the disc 80, can be coated with either conductive material or insulative material. The contacts connected to the four output wires 92 through 98 remain stationary while the disc 80, attached to the X position wheel shaft 99, rotates. Currents flow through the disc and through those wires 92 through 98 which are over a conductive portion of the disc, to indicate position.

In order to indicate many positions, the disc 80 is, in practice divided into a large number of rings and sectors, so that a large number of positions can be indicated and small changes of position are registered. A similar scheme is used for the Y position. The advantage of the readout scheme of FIG. 6 is that a digital output is provided which completely defines the position of the indicator at every instant. A major disadvantage is that a large number of wires must be connected to the position indicator control so that a relatively thick cable trails behind it and limits the ease with which it can be moved.

FIG. 6 illustrates still another position readout means, which possesses the advantage of digital output while requiring a minimum number of leads connecting the position indicator control to the computer. In the readout circuit of FIG. 5, a disc 100 is provided which has three rows of electrical contacts, designated 102, 104 and 106. The disc 100 has its axes fixed to the X wheel of the device shown in FIGS. 1, 2 and 3 in place of the potentiometer. The device of FIG. 6 operates by transmitting "up" pulses when the position indicator control moves to the right and "down" pulses when the position indicator control moves to the left. A counter circuit, which is fixed to the computer adds the up pulses and subtracts the down pulses to provide a continuous digital indication of the position of the position indicator control. A similar arrangement is used for the Y position.

美国
America

3,541,541

5

In the circuit of FIG. 6 three sensors **108**, **110** and **112** are located adjacent to the rows of contacts **102**, **104** and **106**, respectively. Whenever the disc **100** rotates and one of the contacts of one of the three rows comes under a sensor, a voltage is delivered over one of the lines **114**, **116** or **118**, leading to circuitry located at the computer. The contacts on the three rows **102**, **104** and **106** are located so that only one contact is under a sensor at any given time.

The lines **114**, **116**, and **118** trail behind the position indicator control and lead to flip-flop **120**, OR gate **122** and AND gates **124** and **126**, as shown. The outputs of the two AND gates **124** and **126** lead to counter **128** to cause it to count up or down. The counter **128** has numerous output lines over which it continuously transmits signals indicating in a digital manner, the position of the position indicator control.

When the disc **100** of FIG. 6 rotates and a contact of row **104** comes under readying sensor **110**, a signal is transmitted over line **116** to set the RS flip-flop **120**. If the disc is rotating in a forward or clockwise direction the next contact of the three rows to come under a sensor is a contact of up row **102**. When a contact **102** touches sensor **108**, a signal is transmitted over up line **114** to AND gate **124**. Inasmuch as the flip-flop **120** has been set, a voltage is being received at input **130** of gate **124**, and the additional signal at input **132** of gate **124** causes it to generate a pulse. The pulse from gate **124** enters counter **128** at its up input, causing the counter to count up by one digit. The signal over up line **114** also enters OR gate **122** which leads to delay line **124** which, in turn, leads to the reset input of flip-flop **120**. A delay of delay line **124** is very short so that flip-flop **120** is reset an instant after an up pulse is registered by the counter.

If the disc **100** continues to turn clockwise after an up pulse is registered, then a contact of row **106** comes under down sensor **112** and causes it to deliver a signal over down line **118** to the AND gate **126**. The AND gate **126** will not deliver a pulse because its input **134** from the flip-flop **120** has no voltage on it, due to the fact that the flip-flop **120** has been reset. Only after another signal is transmitted over line **116** to again set the flip-flop **120**, will a pulse from up line **114** cause the registation of another up count. Down counting occurs in a similar manner when the disc **100** turns in a counter clockwise direction.

The circuit of FIG. 6 requires only four leads between the computer and the position indicator control. The four leads are the lines **114**, **116**, **118**, and an additional line (not shown) for connection to a voltage source to provide pulses that flow through the sensors to the other three lines.

FIG. 7 illustrates another position readout means of the incremental encoder type, similar to that of FIG. 6. The circuit of FIG. 7 utilizes a simpler encoding disc and one fewer lead, although it involves more complex electronics. In the circuit of FIG. 7, a disc **140** whose axis is connected to a position wheel such as the X position wheel is provided which has a track **142** having spaced conductive segments. A control contact **146** and stepping contact **148** are disposed over the track to make contact with the conductive segments thereof. The contacts are arranged for contacting the segments at angular positions of the disc which overlap. A lead (not shown) connected to the disc **140** conducts currents to the segments of the two tracks.

The contact **146** is connected to a Schmidt trigger circuit **150** which provides currents to two of four AND gates **152**, **154**, **156** and **158**, at a time. The other contact **148** carries current to a resolver **160** of the Schmidt trigger, on-off type, which provides signals with sharp, standard on-off wavefronts. The output of the resolver is delivered to a differentiator **162** which delivers sharp pulses to an inverter chopper **164** and a normal chopper **166**. The outputs of each chopper are delivered to two

6

of the AND gates. The outputs of two of the AND gates deliver pulses to the up input port **168** of an up/down counter **170** while the outputs of two other AND gates are delivered to the down input port **172** of the counter. The counter **170** continuously delivers digital output signals defining the position of disc **140**.

While the position indicator control can be used merely to cause a change in cursor position, and other means such as a typewriter can be used for adding to the pattern, the indicator control can be used in other ways. For example, the position indicator control can be placed on a drawing to be displayed on the CRT, and then the indicator control can be moved to trace the lines of the drawing with the computer causing corresponding lines to be displayed on the CRT. For such uses, the wheels and electrical signal generators of the indicator control should cause cursor movements which very closely correspond to indicator control housing movements.

The particular mechanical construction shown in FIGS. 2 and 3 are especially well adapted for maintaining accuracy of output and ease of use. The use of only three points of contact, comprising the two wheels and the ball bearing support, help to assure that both wheels will constantly remain in firm contact with the surface on which the position indicator control rests. The location of the various buttons for indicating areas of the display to be operated on, or for other purposes, on the indicator which is moved by the hand allows a human operator to maintain control over both position of changes and the type of changes on the display with only one hand. The use of an indicator control which rests firmly on a surface enables the operator to accurately maintain position with a minimum of muscle effort, since the indicator control remains stationary unless some force is applied to it. The use of relatively large position wheels having appreciable, even if small, moments of inertia, reduces jittering of the indicator control and promotes smooth movement which is helpful in accurate positioning where the displayed characters are small or where accurate tracing of a pattern is required.

While particular embodiments of the invention have been illustrated and described, it should be understood that many modifications and variations may be resorted to by those skilled in the art, and the scope of the invention is limited only by a just interpretation of the following claims.

I claim:

1. In a display system controlled by a computer whereby the display is alterable in accordance with signals delivered to said computer which indicate positions on said display and changes desired to be made therein, the improvement in a position indicating control apparatus which is movable over a surface to provide position indications corresponding to positions on said display comprising:
   a housing;
   a first position wheel rotatably mounted on said housing and having a rim portion extending past the boundaries defined by said housing for supporting said housing on said surface;
   a second position wheel rotatably mounted on said housing with its axis of rotation oriented perpendicular to the axis of said first wheel, said second position wheel having a rim portion extending past said housing for supporting said housing on said surface;
   transducer means connected to each of said first and second wheels, for generating digital position indicating signals indicating the degree of rotation of said wheels; and
   flexible conductor means for connecting said transducer means to said computer, for conducting said position indicating signals to said computer while enabling unrestrained movement of said housing relative to said computer.

3,541,541

7

2. The improvement in a position indicating control apparatus as defined in claim 1 wherein:
   said transducer means comprises an incremental encoder connected to said first position wheel and said flexible conductor means for generating first pulses at each predetermined increment of rotation of said first position wheel in a first direction and for generating second pulses at each increment of rotation of said first position wheel in the opposite direction; and including:
   counter means connected to said flexible conductor means, for generating a digital count indicating the net rotation of said first position wheel.
3. The improvement in a position indicating control apparatus as defined in claim 2 wherein said incremental encoder comprises:
   a disc connected to said first position wheel having track means, said track means having a plurality of spaced conductor segments;
   a control contact and a stepping contact disposed along said track means, said control and stepping contacts positioned for the electrical connection of only one of said contacts with said segments at first predetermined angular positions of said disc and for the simultaneous electrical connection of both of said contacts with said segments at second predetermined angular positions of said disc; and
   logic means connected to said control and stepping contacts for generating said first pulses when said stepping contact makes a first direction of transition between electrical connection with one of said segments and lack of electrical connection with one of said segments at the same time that said control contact is in a first predetermined state of electrical connection with one of said segments and for generating said second pulses.
4. The improvement of a position indicating control apparatus as defined in claim 1 wherein:
   said transducer means comprises a shaft position encoder having a plurality of outputs and said conductor means comprises a plurality of conductors connected to said outputs of said encoder, whereby to constantly indicate the position of said position indicating apparatus.
5. In a display system controlled by a computer whereby the display is alterable in accordance with signals delivered to said computer which indicate positions on said display and changes desired to be made therein, the improvement in a position indicating control apparatus which is movable over a surface to provide position indications corresponding to positions on said display comprising:
   a housing;
   a first position wheel rotatably mounted on said housing and having a rim portion extending past the boundaries defined by said housing for supporting said housing on said surface;

8

   a second position wheel rotatably mounted on said housing with its axis of rotation oriented perpendicular to the axis of said first wheel, said second position wheel having a rim portion extending past said housing for supporting said housing on said surface; and
   transducer means connected to each of said first and second wheels, for generating digital position indicating signals indicating the degree of rotation of said wheels.
6. The improvement described in claim 5 including:
   coupling means for substantially unrestrained coupling of said transducer means to said computer, to couple said position indicating signals to said computer while enabling substantially unrestrained movement of said housing relative to said computer.
7. The improvement described in claim 5 including:
   a flexible conductor for connecting said transducer means to said computer, to carry position indicating signals to said computer while enabling substantially unrestrained movement of said housing relative to said computer.
8. A display system for presenting an alterable visual display comprising:
   cathode ray tube means for providing a visual display;
   computer means connected to said cathode ray tube means for controlling inputs to said cathode ray tube means to define the visual display thereof, said computer means including means for generating signals defining a cursor for display at variable positions on said cathode ray tube means and means for altering inputs to said tube means to cause a change in the display about the position of said cursor;
   a position indicator control connected to said computer means, said position indicator control having a housing which contains transducer means for delivering signals for causing movement of said cursor on said cathode ray tube means in response to movement of said housing over a surface; and
   at least one cathode ray tube display control switch disposed on said position indicator control.

References Cited

UNITED STATES PATENTS

3,304,434	2/1967	Koster	33—141.5 X
3,346,853	10/1967	Koster et al.	340—324 X
3,355,730	11/1967	Neasham	178—18 X

THOMAS B. HABECKER, Primary Examiner

D. L. TRAFTON, Assistant Examiner

U.S. Cl. X.R.

33—141.5; 178—18; 345—177, 204, 212, 354

CT扫描仪的发明

CT Scanner invention

G. N. 亨斯费尔德

Godfrey Newbold Hounsfield

1895年，德国物理学家W.K.伦琴在进行克鲁克斯管实验时发现了χ射线（χ-ray）。1913年，W.D.库利吉发明了真空χ射线管。χ射线具有很强的穿透力，使现代医学对健康和疾病的诊断迈向无创伤、无侵犯的新时代。

1968年8月23日，英国人G.N.亨斯费尔德在英国申请了名称为“多角度测量χ射线或γ射线吸收，透过情况和数据分析的方法和仪器”的专利，即首次在专利申请中揭示了CT技术。这项发明，引起了科技界的极大震动，被誉为自伦琴发现χ射线后，放射诊断领域最重大的成就。

US 3778614

www.sipo.gov.cn

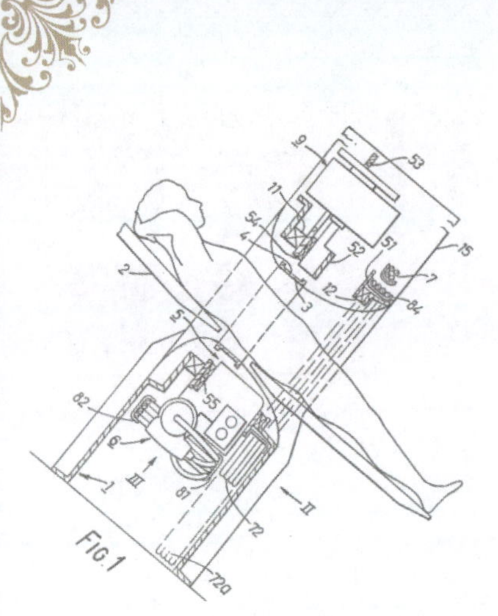

第十八篇

CT 扫描仪

CT（Computed Tomography），即电子计算机断层扫描技术，它利用精确准直的 χ 线束、γ 射线、超声波等，与灵敏度极高的探测器一起围绕人体的某一部位做一个接一个的断层扫描，根据人体不同组织对 χ 线束的吸收与透过率的不同，将测量所获取的数据输入电子计算机，电子计算机对数据进行处理后，就可摄下人体被检查部位的断面或立体的图像，从而为医生诊断病情提供影像学参考。那么第一台 CT 扫描仪是怎么发明的呢？

自从 χ 射线被发现后，医学上就开始用它来探测人体疾病。但是，由于人体内有些器官对 χ 射线的吸收差别极小，因此 χ 射线对那些前后重叠的组织的病变就难以发现。于是，美国与英国的科学家开始寻找一种新的东西来弥补用 χ 射线技术检查人体病变的不足。

1963 年，美国物理学家阿伦·考马克（Allan Cormack）发现人体不同的组织对 χ 射线的透过率有所不同，在研究中还得出了一些有关的计算公式，这些公式为后来 CT 的应用奠定了理论基础。

1967 年，英国电子工程师 G. N. 亨斯费尔德（Godfrey Newbold Hounsfield，1919–2004 年）于 1919 年出生在英国的诺丁汉（Nottinghamshire），在农场长大，喜欢做科学实验。“二战”期间，他从事雷达工作，“二战”结束后进入大学学习。1951 年亨斯费尔德进入 EMI 有限公司（Electrical and Music Industries）工作，从事计算机

领域研究。有一次一个研究项目的夭折使他重新开始思考下一步的研究方向，甚至在工作之余也没有停止思考这个问题，渐渐地他的思路清晰起来，要使计算机能实现模式识别，再通过这些模式能“读”数据和数字，就要把这些模式识别和雷达原理结合起来，并建立有关计算方法。经过一段时间努力，他终于完成了一个发明构思，用 χ 射线来多角度产生人体薄片层面透过数据，并用计算机对这些数据进行加工处理，以便最终重建立体图像。然后他制作了一台能加强 χ 射线放射源的简单的扫描装置，即后来的 CT，用于对人的头部进行实验性扫描测量。后来，他又用这种装置去测量全身，获得了同样的效果。

1968 年 8 月 23 日，亨斯费尔德将这一发明在英国申请了专利，并于 1972 年获得名为“用 χ 射线或 γ 射线吸收检查人体的方法和仪器”的英国 GB1283915 号专利权，如图 18-1、图 18-2 所示。

PATENT SPECIFICATION 1283915

图 18-1

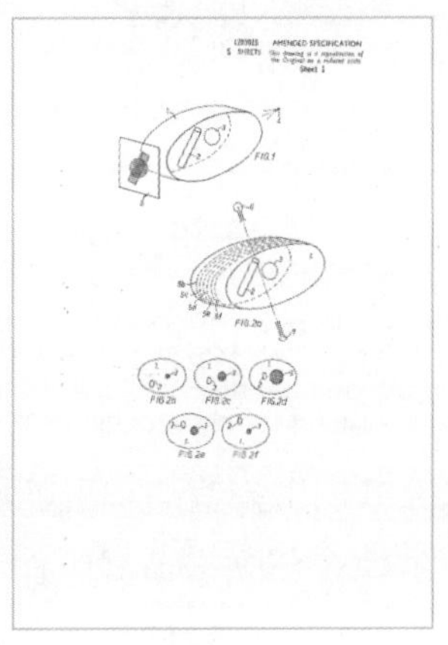

AMENDED SPECIFICATION

图 18-2

从专利文献库中我们发现，亨斯费尔德在完成英国专利申请后能很好地运用英国优先申请及美国专利制度，将此发明在引入美国市场时即利用了英国申请有效期的最后两天，于 1969 年 8 月 21 日先向美国提交了 No.861538 申请案，后又运用美国“继续申请”制度在对技术方案进行了一定的修改后，于 1971 年再次提交了 No.861538 申请案的继续申请，并于 1973 年获得名为“多角度测量 χ 或 γ 射线吸收，转换情况并数据分析的方法和仪器”的美国 US3778614 号专利

权，如图 18-3、图 18-4 所示，其专利权人为 EMI 有限公司。同时，通过同族专利检索发现，亨斯费尔德除在英国本国申请以外，还在美国、日本、法国、德国、荷兰、加拿大等国家都申请了专利权。另外，我们还发现在此发明基础上有 170 多位后来发明者在其专利申请中都参考了该专利，从这两个角度可以反映此项发明在其领域内的技术价值。

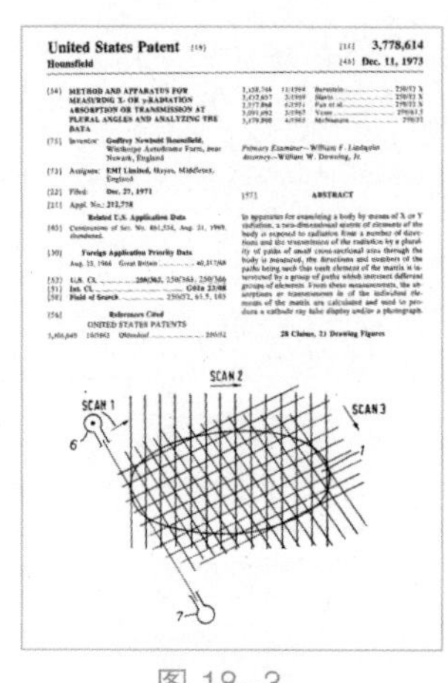

United States Patent [19]
Hounsfield

[11] 3,778,614
[45] Dec. 11, 1973

[54] METHOD AND APPARATUS FOR MEASURING X- OR γ-RADIATION ABSORPTION OR TRANSMISSION AT PLURAL ANGLES AND ANALYZING THE DATA

[73] Assignee: EMI Limited, Hayes, Middlesex, England

[22] Filed: Dec. 27, 1971

Related U.S. Application Data

Foreign Application Priority Data

[56] References Cited
UNITED STATES PATENTS

[57] ABSTRACT

28 Claims, 21 Drawing Figures

图 18-3

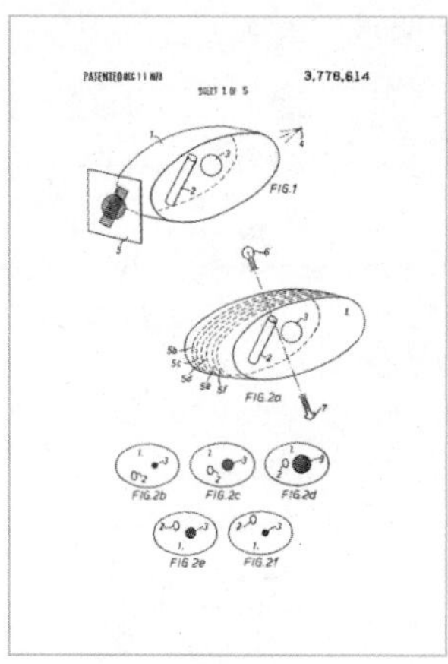

图 18-4

从专利文献中我们发现，在亨斯费尔德的这件美国 US3778614 号专利申请提交时已有 6 件美国在先申请，其中一件名为“用于研究物体内部选定区域稠密物质的辐射能设备”的美国 US3106640 号专利，正是亨斯费尔德在英国申请案中作为发明背景技术的在先专利。该技术方案的一个缺陷是，被检查者的身体必须多次通过光束，以在显示大小的一个二维区域内建立起由两个维度指示的吸收（或传输）系数元素组成的数模，才能解读被检查者身体有关部位的情况。

而亨斯费尔德的发明，则是通过利用 χ 或 γ 射线照射检查人体的仪器，从若干方向对人体的两维矩阵单元进行照射，并且用多路径测量穿越人体小层面照射的透过情况，路径的方向和数量使得通过层面不同组群单元的一组路径要横断每个矩阵单元。用这样的方法检查连续的平行薄片层面，所产生薄片层面照片能组合成较大部分或整个

的人体照片，将其存储在诸如磁带等记录器中，计算机运算处理将人体每 3mm 层体立方体的数字量表示其材料的 χ 射线吸收率，再将其通过数模转换器变成模拟型式，把数字矩阵中的每个数字转换成由黑到白不等灰度的小方块，这样就能发现人体内有关部位的细小病变。

为了使自己的发明能够应用于临床，1971 年 9 月，亨斯费尔德又与一位神经放射学家合作，在伦敦郊外一家医院安装了他设计制造的这种装置，开始了头部检查。1971 年 10 月 4 日，这家医院用该装置检查了第一个病人。该 CT 扫描仪通过 χ 射线束绕头部可转 180 度，每个视角都可照射，然后合成图片。患者在完全清醒的情况下朝天仰卧，χ 射线管装在患者的上方，绕检查部位转动，同时在患者下方装一计数器，使人体各部位对 χ 射线吸收的多少反映在计数器上，再经过电子计算机的处理，使人体各部位的图像从荧屏上显示出来。这次试验非常成功，效果很好。

1972 年 4 月，亨斯费尔德在英国放射学年会上首次公布了这一结果，正式宣告了 CT 扫描仪的诞生。这一伟大的发明引起了医学界极大的震动，被誉为自伦琴发现 χ 射线后，放射诊断领域最重大的成就。为此，亨斯费尔德与美国物理学家阿伦·考马克（Allan Cormack）共同获得 1979 年诺贝尔医学奖。

亨斯费尔德所在的 EMI 有限公司 1974 年研制出全身 CT 扫描仪，检查范围扩大到胸、腹、脊柱及四肢。1975 年，改进型的 CT 扫描仪研制成功，该仪器带有可将人体送入圆形孔检查的可移动床，可进行全身扫描。我们从专利文献库中也能找到其身影。1976 年，EMI 有限公司在英国以“χ 射线扫描仪的连接装置”为名提出一件申请，并于 1977 年获得英国 GB1536427 号专利权，如图 18-5、图 18-6 所示。

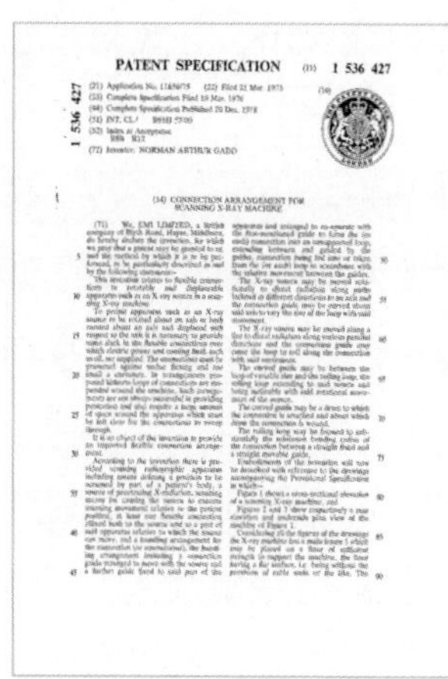

PATENT SPECIFICATION (11) 1 536 427

(72) Inventor: NORMAN ARTHUR GADD

(54) CONNECTION ARRANGEMENT FOR SCANNING X-RAY MACHINE

(71) We, EMI LIMITED, a British company of Blyth Road, Hayes, Middlesex, do hereby declare the invention, for which we pray that a patent may be granted to us, and the method by which it is to be performed, to be particularly described in and by the following statement:—

图 18-5

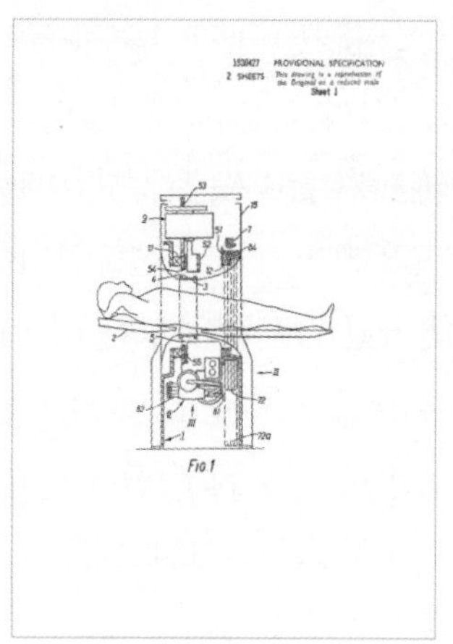

图 18-6

这些技术改进为现代化的 CT 扫描仪奠定了基础。EMI 有限公司有了从乐器唱片转向扫描仪发展的财力支持，但由于销售不好，在 1980 年将扫描仪的专利权转让给了美国通用电气公司。

现今，CT 作为一种计算机层析成像技术，在医学、工业、地球物理、农业、工程检测和探测等多方面发挥着越来越重要的作用，随着科学技术的进一步发展，CT 技术将向着多源、多排、多层方向发展，以求得扫描速度、覆盖范围、图像质量的同时改善。同时便携化的 CT 技术也将成为今后 CT 发展的方向。

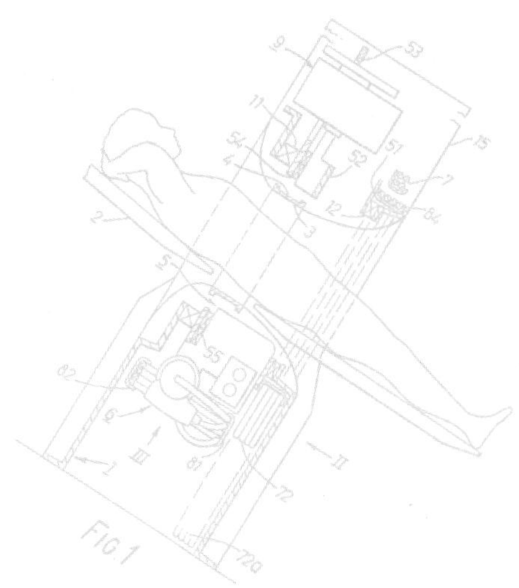

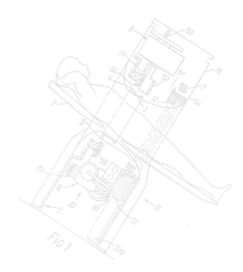

1973. US 3778614

多角度测量 χ 或 γ 射线吸收，转换情况并数据分析的方法和仪器

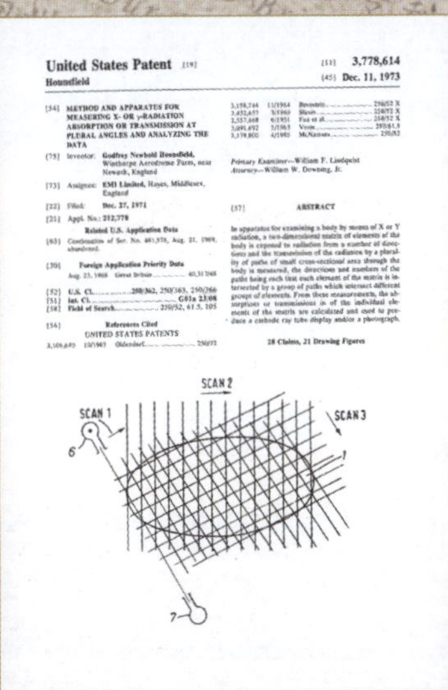

United States Patent [19]
Hounsfield

[11] 3,778,614
[45] Dec. 11, 1973

[54] METHOD AND APPARATUS FOR MEASURING X- OR γ-RADIATION ABSORPTION OR TRANSMISSION AT PLURAL ANGLES AND ANALYZING THE DATA

[75] Inventor: Godfrey Newbold Hounsfield, Winthorpe Aerodrome Farm, near Newark, England

[73] Assignee: EMI Limited, Hayes, Middlesex, England

[22] Filed: Dec. 27, 1971

[21] Appl. No.: 212,778

Related U.S. Application Data

[30] Foreign Application Priority Data

[52] U.S. Cl.
[51] Int. Cl. G01n 23/08
[58] Field of Search

[56] References Cited
UNITED STATES PATENTS

Primary Examiner—William F. Lindquist
Attorney—William W. Downing, Jr.

[57] ABSTRACT

In apparatus for examining a body by means of X or Y radiation, a two-dimensional matrix of elements of the body is exposed to radiation from a number of directions and the transmission of the radiation by a plurality of paths of small cross-sectional area through the body is measured, the directions and numbers of the paths being such that each element of the matrix is intersected by a group of paths which intersect different groups of elements. From these measurements, the absorptions or transmissions of the individual elements of the matrix are calculated and used to produce a cathode ray tube display and/or a photograph.

28 Claims, 21 Drawing Figures

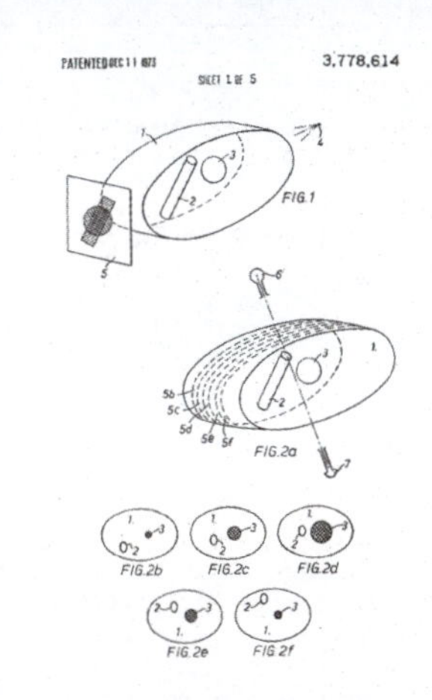

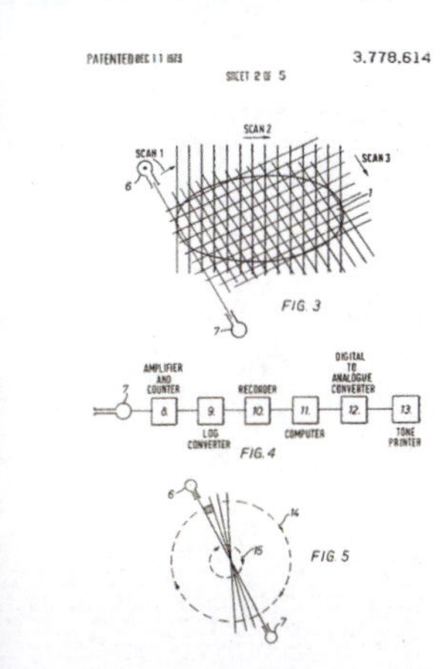

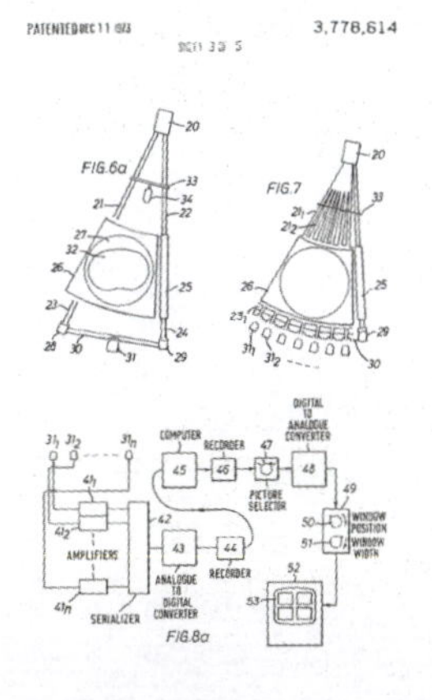

美国 America

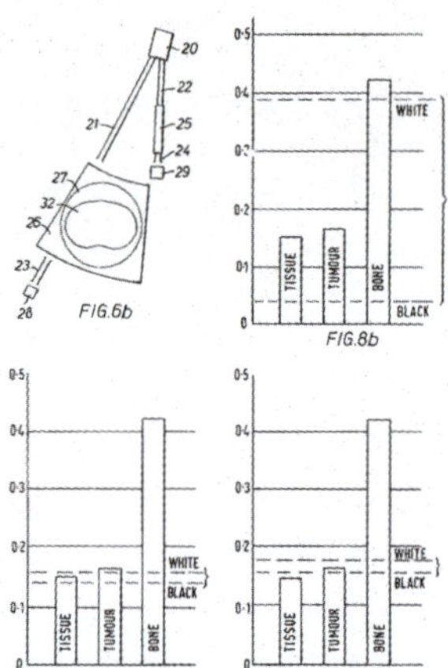

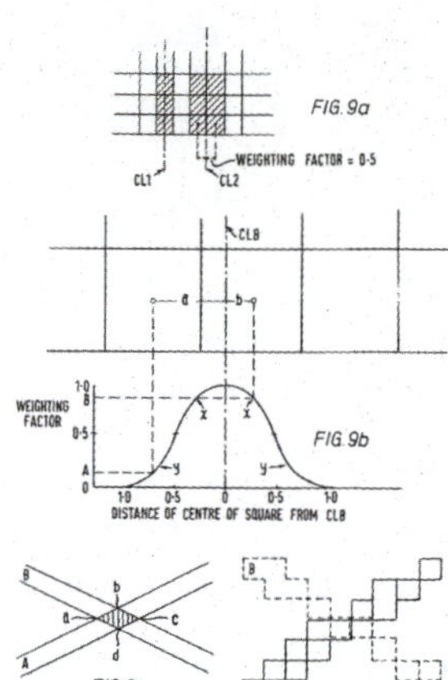

1

METHOD AND APPARATUS FOR MEASURING X- OR γ-RADIATION ABSORPTION OR TRANSMISSION AT PLURAL ANGLES AND ANALYZING THE DATA

This is a continuation of application Ser. No. 861,538, filed Aug. 21, 1969, now abandoned.

This invention relates to a method of, and apparatus for examining a body by means of radiation such as X or γ radiation.

The method and apparatus according to the invention can be used to produce radiographs in any convenient form, such as a picture on a cathode ray tube or other image forming device, a photograph of such a picture, or a map of absorption coefficients such as may be produced by a digital computer and on which "contours" may subsequently be drawn.

It is well known that when an X-ray picture is taken through an object, the three-dimensional interior appears only as a two-dimensional picture, the details from front to back of the object appearing superimposed and the picture being consequently confused and difficult to interpret. For example, if the object were a book, a conventional X-ray picture would reveal little about the contents, as the information on any one page could not be extracted from the superimposed information from all the other pages. A method of carrying out examination of a body by X-radiation known as tomography has been proposed with a view to reducing this disadvantage. In this method, a source of radiation is orbited relative to the body being examined, about an axis in or near the body, while a plate or other screen sensitive to radiation transmitted through the body is displaced in such a way that the elements in one plane of the body remain substantially stationary. This method has however the disadvantage that the shadows of elements on other planes of the body move and this movement not only obscures the information about the selected plane, but results in a loss of information about the other planes.

It has also been proposed to map the absorption coefficient of a two dimensional slice of a body from a knowledge of the line integral of the absorption coefficient along all lines intersecting the slice, by a process involving the application of Fourier inverting techniques. This proposal is described in a paper entitled "Representation of a Function by its Line Integrals, with some Radiological Applications" by A.M. Cormack (Journal of Applied Physics, Volume 14, Number 9, pages 2722 to 2727 and Number 10, pages 2908 to 2913). An experimental test was carried out on a special model, each line integral used in the evaluation being derived by letting a fine beam of γ-rays of known intensity be incident on the slice and deriving a signal representing the integral along the line of the beam from the γ-rays emerging from the body. The method described in this paper is capable in theory of yielding a unique principal solution, but is nevertheless complicated, has limited practical application and liable to error in the practically feasible forms.

It has also been proposed in a paper entitled "Transmission Scanning: A Useful Adjunct to Conventional Emission Scanning for Accurately Keying Isotope Deposition to Radiographic Anatomy" by D.E. Kuhl M.D., John Hall Ph.D., and W.L. Eaton M.D. (Radiology, 1966, Volume 87, pages 278 to 284) to scan a section of a body by moving a small radio active source of

2

either Am^{241} or I^{131} so as to follow the motion of a detector at the other side of the body. It is understood that a representation of the section similar to a roentgenogram was produced by causing a thin line of light to be generated across the face of an open-shutter oscilloscope to represent the line of view of the detector. The line of light was moved across the oscilloscope face to correspond to the detector motion, and the film holder rotated step-wise to correspond with the angle of the detector. The brightness of the line of light was varied according to the counting rate from the detectors so that the film ultimately recorded a picture built up from a series of overlapping lines of varying brightness. This method was applied only to a relatively small number of angles of view and it will be appreciated that the density of each picture point was affected by all other picture points traversed by the beam.

One object of the present invention is to provide a relatively simple method and apparatus for examining a body by means of penetrating radiation which is able to detect quite small differences in the transmissions or absorptions of element in a planar slice of the body.

Another object of the invention is to provide an improved method of, and apparatus for, examining a body by means of penetrating radiation which is able to provide an accurate representation of elements of the body without the need to administer radio opaque contrast media as in angiography.

Another object of the present invention is to provide an improved method and apparatus whereby a radiographer, working unaided, can obtain representations of sectional slices of a body relatively quickly.

Another object of the present invention is to provide an improved method of, and apparatus for, producing representations of cross sectional slices of a body which substantially avoids the disadvantages of the attenuating effects of bone and other matter, which can operate with high resolution, and which reduces disadvantages due to scatter.

With these and further objects in view, the invention provides, in general terms, a method of examining at least part of the interior of a body using penetrating radiation such as X or γ rays, comprising the steps of:

a. transmitting radiation from an external source through the body in a plurality of rays traversing a plurality of respective paths at an initial angle or initial mean angle, said rays constituting a first set of rays in a single planar slice of said body, the cross-sectional dimensions of each ray being small in relation to the dimensions of the body;

b. transmitting radiation from said external source through the body in further sets of rays similar to said first set of rays, said further sets being disposed in said planar slice at angles or mean angles different from each other and from said initial angle or initial mean angle;

c. the sets of rays being such that every element of a two-dimensional matrix of elements of the body in said planar slice is intersected by a group of said rays, the group of rays being different for the different elements of the array;

d. deriving from each ray emerging from the body an output signal representing the sum of the transmissions or absorptions in the elements of the body intersected by the ray, thereby to derive sets of output signals corresponding to the sets of rays sufficient to obtain the

3

transmission or absorption of each element of said matrix;

e. deriving from said output signals, by a process of successive approximations, resultant signals representing the transmissions or absorptions of the elements of said matrix, the derivation of said resultant signals including the steps of

i. deriving a difference signal responsive to the difference between each output signal and its reconstruction from the last approximations to said resultant signals, and

ii. adjusting the approximations to the respective resultant signals in response to said difference signal,

f. producing in response to said resultant signals a representation of the transmissions or absorptions of said elements of the slice of the body.

The present invention also provides apparatus capable of being used for carrying out the aforesaid inventive method of examining a body

In order that the invention may be clearly understood and readily carried into effect, the same will now be described with reference to the accompanying drawings in which:

FIG. 1 shows the kind of picture produced by conventional X-ray apparatus,

FIGS. 2a, 2b, 2c, 2d, 2e, and 2f illustrates the principle of the invention and the kind of picture produced by the invention,

FIG. 3 shows one method of scanning used in the invention,

FIG. 4 shows in block form suitable apparatus for carrying out the invention,

FIG. 5 shows an alternative method of scanning,

FIGS. 6a and 6b illustrates diagrammatically the construction of the scanning means of apparatus according to two other examples of the invention,

FIG. 7 illustrates a modification of FIG. 6a,

FIG. 8a illustrates, partly in block form, the apparatus embodying the scanning means illustrated in FIG. 7,

FIGS. 8b, 8c and 8d are diagrams useful in explaining the operation of parts of the apparatus shown in FIG. 8a, and

FIGS. 9a, 9b, 9c and 9d illustrates the application of weighting factors to elements of the picture.

Referring to FIG. 1, this shows a body 1 containing a bone 2 and a tumour 3. Also shown are a source of X-rays 4 and a X-ray film 5. As can be seen, images of the bone and tumour are produced on the film, but partly superimposed. The tone of any point on the film is dependent on the product of the co-efficients of transmission of all the elements lying between that point and the X-ray source. Thus if the bone 2 has the lowest co-efficient of transmission, the tumour 3 the second lowest and the surrounding material the highest, the X-ray image comprises a dark patch where the bone and tumour are superimposed, a lighter patch due to the bone not superimposed on the tumour and a still lighter patch due to the tumour not superimposed on the bone. These are surrounded by a light area where neither bone nor tumour is present. Also has the differences between the coefficients of transmission of tumour and normal tissue are small, the differences in tone between the different parts of the X-ray picture are slight and difficult to detect using such a method.

Referring now to FIG. 2a, the body, bone and tumour are denoted by the same references as in FIG. 1. The

4

X-ray source is replaced by a source 6 which may also be of gamma rays but is preferably of X-rays. It differs from source 4 in that it produces a beam of small cross section area, for example 3m.m square or diameter, and preferably includes a collimator to reduce scatter of the rays. The X-ray film 5 has been replaced by a detector 7, which may be a scintillator and a scintillation counter and which preferably also includes a collimator. The body 1 is scanned by the beam in one plane only, the plane being 3m.m. thick in this example, in a direction not only linearly across the plane, but at a plurality of angles round the plane, the detector 7 being so mounted that it is always pointing towards the source 6. FIG. 3 illustrates the scanning in more detail. If only a single scan across the plane were performed, the result would merely be equivalent to a conventional X-ray picture of that plane, all the objects on a line between source 6 and detector 7 being superimposed. However by performing a large number of scans, sufficient information can be derived to enable the coefficient of absorption of the material in each 3m.m. cube of material in the plane to be calculated and the coordinates of its position in the plane determined. Although only three scans are shown in FIG. 3, it will be appreciated that many more would be required in practice.

In each position of the beam the detector 7 determines the transmission of the X radiation by a path of relatively small cross-sectional area through the body. The plane under examination is regarded as a two dimensional matrix of elements and the directions and numbers of the paths is such that each element of the matrix is intersected by a group of paths, which paths intersect different groups of elements.

From the transmissions by all the paths, a series of simultaneous equations is built up represented by the discrete output signals derived from the radiation traversing all the respective paths and the solution of these equations by means of a digital computer provides the absorption coefficient of each element of the matrix. The outputs of the computer may be used to produce a picture or other representation of the section in any convenient form. Successive parallel planes may be examined in this way, and a picture of each planar slice produced to build up a picture of the entire body or a larger section of it. The slices may be examined in sequence or simultaneously by using a number of X-ray sources and detectors in parallel. FIGS. 2b to 2f show the pictures resulting from examination of planar slices 5b to 5f of body 1.

FIG. 4 shows a block diagram of the apparatus for producing pictures from the outputs from detector 7. The output from detector 7 is applied to an amplifier and counter 8 which produces a digital output representing the number of counts in each reading. The output from 8 is converted to logarithmic form in a logarithmic converter 9 whose couput is stored in a punched-tape or magnetic tape recorder 10 before being transmitted to a digital computer 11 for processing. The computer 11 produces for each 3m.m. cube of a planar slice of body 1 a digital number representing the absorption coefficient of the material within that cube. These digital numbers may be converted to analogue form in digital-to-analogue converter 12 and applied to a tone printer 13 to produce a picture. Alternatively, the computer outputs may be retained in digital

美国 America

form for comparison by pattern recognition techniques, with other digitised pictures.

To achieve the required result, the absorption along each path is deduced from the transmission by each path and a knowledge of the initial intensity of the beam or ray entering each path. The logarithmic converter **9** is used to provide a linear output so that the total absorption along a path is equal to the sum of the absorption in each small element along the path. Let 100 parallel paths be used for each of 400 directions spaced equally over 180°. The computer **11** has then 40,000 figures to process, each representing the total absorption along a given path. Consider the section divided into 100 × 100 similar meshes as on Cartesian graph paper. Each mesh represents an element of the body, but the term mesh will be used in the following mathematical consideration for convenience. The computer **11** is then programmed to give the absorption for each of the 10,000 meshes.

Consider a ray which passes through a set of $n=100$ meshes through none of which a ray has previously passed. Let the total absorption be Z dB. The computer then allocates a provisional value of Z/100 to each of the meshes. Now suppose that, at a later stage, a ray passes through another set of 100 meshes the absorption in some or all of these meshes having already been allocated. Let the sum of the figures already allocated be Z_1 whereas the new measurement gives a total absorption Z_2. It will be appreciated that Z_1 constitutes a reconstruction of the output signal Z_2 derived from the last approximation to the absorptions of the respective meshes. Then a correction $(Z_2-Z_1)/100$ is added to the figures already appearing in each of the meshes. This process is then continued as successive approximations for all the 40,000 rays. This process gives a rough approximation, but to obtain better accuracy, the computer must repeat it a number of times, say five.

Consider a single section of the body in the xy-plane in which the absorption per unit distance in dB at the point x, y is z. Let $z=f(x,y)$. Now consider a single set of rays all parallel to the y-axis and spaced equally by intervals Δx. The rays are arranged to have a width rather greater than Δx so that some overlapping occurs. The optimum beam width is determined empirically. For mathematical purposes it is assumed that the change of absorption through any interval Δx to be negligible. We now suppose that the section of the body to be examined is bounded on two sides by the x- and y-axes and is square in shape so that it can be divided into elementary squares with edges parallel to the axes.

The complete total of rays can be divided into sets each of which consists of parallel rays or effectively parallel rays at a given angle or mean angle. The sets of discrete output signals derived from the rays in each set are treated in the computer in sequence. However, since there are only about 100 × 100 meshes and about 400 angles are employed within 180°, rays at neighbouring angles must include some of the same squares and their absorption will not, therefore, be independent. If the sets were therefore taken in angular succession the lack of independence would clearly lead to a slower convergence than if they were independent.

The computer is therefore arranged, by programming, to take the different angular sets in a pseudo random order with large angular gaps, of say 40°, between successive sets of rays. The sequence is intended to ensure that every angle is included, but not repeated, within the 400 directions. Rays close together in angle then appear far apart in the computer scanning sequence.

The accuracy of detection by detector such as **7** is limited so that the raw data contain errors and therefore, as the complete cycle of 100 × 400 measurements is analysed 4 or 5 times by the computer, the resultant figures for the meshes tend to oscillate. It has been found that this can be avoided by multiplying the later corrections by a factor which is less than unity and falls steadily for successive cycles.

The procedure may be represented mathematically as follows. The true continuous distribution function is given by equation (1)

$$z=f(x,y)\ (1)$$

Let the distribution function reached at some stage of the work be

$$z'=g(x,y) \tag{2}$$

which is a discontinuous function since z' must have the same value over each mesh.

Now consider an arbitrary ray passing through n meshes. If z_r is the mean value of z through the r th mesh and Z is the total absorption (or attenuation) of the ray in dB as measured

$$Z=\sum_1^n z_r \tag{3}$$

The value of z' for each mesh will also be known from the previous work. If no ray has passed through a given mesh z' is put equal to zero.

The mean square error for all the meshes along the path of the ray will be written E where

$$nE=\sum_1^n (z'_r-z_r)^2 \tag{4}$$

and it is required to choose new values z''_r to replace z'_r in order to minimise E.

There is no reason to favour one mesh rather than another, and therefore a constant C is added to z'_r where C is independent of r and must be obtained from the additional information provided by Z. Hence

$$z''_r=z'_r+C \tag{5}$$

Hence the new value of n E will be

$$nE=\sum_1^n(z''_r-z_r)^2=\sum_1^n(z'_r-z_r+C)^2 \tag{6}$$

The minimum value of E is obtained where C is equal to the mean value of z_r-z_r' or

$$C=\frac{\sum_1^n (z_r-z'_r)}{n}=Z/n-\frac{\sum_1^n z'_r}{n}$$

Hence

$$z''_r=z'_r+Z/n-\frac{\sum_1^n z'_r}{n}$$

In other words the correction applied is equal to the mean value of the error. If none of the n meshes has previously appeared all the z'_r are put equal to zero so that

$$z''_r=Z/n \tag{9}$$

In other words the attenuation is, at first, uniformly distributed among the meshes.

For the s th ray equation (**8**) becomes

$$z''_{rs}=z'_{rs}+Z_s/n-\frac{\sum_{r=1}^n z'_{rs}}{n}$$

If there is a total of S rays there will be a total of S equations for a complete cycle. If m is the number of rays in a set of parallel rays and N is the number of angles

$$S=m\,N \tag{11}$$

If q is the number of complete cycles used by the computer, the total number of ray operations is q S.

Since the number of rays S per cycle is several times the number of meshes M, the number S of equations will be several times the number M of independent equations.

Difficulties arise in finding a system which traces through the picture matrix an equivalent beam or ray as it has been called heretofore which has effectively constant width, and which also includes the correct number of picture elements along its length, both of these requirements are essential for the accurate computer calculations which are to follow.

The two worst cases are shown in FIG. 9*a*, where in one case a beam centre line CL1 passes through the squares of the matrix perpendicularly and the centre line of the beam passes through the centre of the squares; in the other case the beam centre line CL2 passes between the squares. The latter case would add up to twice as many squares as the former, when the squares along the length of the beam are added up, and would clearly give an error of 2:1.

In order to overcome the above problem the values in each square are multiplied by a weighting factor which is a function of the distance from the centre of the square to the centre line of the beam, i.e., the squares of beam **2** in FIG. 9*a* would have a weighting factor of 0.5, the resulting sum of the numbers in the two beams then being equal.

FIG. 9*b* shows an intermediate position of the beam in which the distances from the centre line CLB of the beam to the centres of the two affected squares in the beam are a and b respectively. The corresponding weighting factors A and B can be read off the graph, and when these are added together they must for reasons indicated above add up to unity. Therefore it follows that the parts of the curves labelled x must be drawn the inverse of the parts labelled y, if the beam and hence the weighting curve is to be considered symmetrical about its centre line.

It can be shown that one requirement for accurate summation of values of the matrix squares is idealised in FIG. 9*c* and its practical equivalent is shown in FIG. 9*d* using a matrix with a beam at the same angle.

In FIG. 9*c* the area abcd is obviously constant at any position of the intersection of the beams and is a function of the angle of intersection of the two beams A and B. In FIG. 9*d* the two equivalent beams vary in width from one to two squares and a constant area at intersection would be impossible without the use of weighting factors. It can be shown that for a given X-ray beam width there is one weighting curve which fulfils all the requirements. For example, if the squares contained at the intersection of the beams in FIG. 9*d* are multiplied by their respective weighting factors taken from this curve, they will produce a sum which is proportional to the area abcd in FIG. 9*c*. Any angle of intersection may be chosen and the beam in FIG. 9*b* may be intersected anywhere along its length for this condition to remain true.

The weighting factor curve can be split up into a table of approximately 20 values to which the computer can refer during calculations without substantially impairing the accuracy of the system.

In the example illustrated in FIG. 3 only a single detector **7** is indicated. If however a fan-shaped or strip beam of radiation is used, with a group of detectors each for receiving radiation transmitted by one of the paths of small cross sectional area, some correction may have to be made in solving the equations for the effect of Compton scatter but in many cases this can be avoided by adequate spacing of the detectors.

As was previously mentioned, the differences in absorption between different materials is very small. However, in accordance with the invention the contrast of the picture produced can be so arranged that the full black to white range represents only the small range of absorption values which is of interest.

It is essential in all X-ray apparatus to ensure that the patient does not receive an overdose of radiation. In this respect the use of a scintillator and a scintillation counter is advantageous as its efficiency and accuracy in detecting X-rays are several orders better than those of photographic film. The maximum detail obtainable in a picture is a function of the number of counts per reading received by the scintillation counter around the edge of the body. In view of the limitation on the permissible number of counts per reading, it would not be feasible to produce a picture having the same order of definition as a television picture when examining a living body, although a high definition picture of an inanimate object could be produced. Moreover, in examining living bodies, it is not normally necessary to have a high definition picture of the whole body. Apparatus according to the invention can be used to produce a picture which is of high definition in the area of immediate interest and of low definition in surrounding areas. For example, as shown in FIG. 5 the radiation source **6** and detector **7** may be arranged to perform a circular scan indicated by the arrow **15** round the edged of the body, which is so positioned that the area of interest is near the centre of the scan. By averaging the number of counts over a small angle of rotation, mean values of absorption for areas enclosed by the angle such as the area shown shaded, may be calculated. It is clear from FIG. 5 that near the edges of the body only a relatively small number of large area elements are being examined, whereas at the centre a large number of small area elements is examined. Consequently the resulting picture will have a high definition near the centre and a low definition towards the

edges. In producing the picture, the points may conveniently be plotted in polar co-ordinates. As in the example of FIG. 3, a large number of scans is required to produce sufficient information. In the embodiment of FIG. 5, the additional scans may be produced by superimposing a slower rotary motion which shifts the axis of the main rotation so that the centre of the circle of the main scanning motion traces a circle of small diameter. This additional scanning motion produces the intersecting paths for each element of the matrix according to which the body is examined. The superimposed motion need not be circular and need not be confined to the centre circle. For example it could be a spiral starting at the edge of the outer circle progressing rapidly towards the centre then performing a slow spiral in the region of the centre. It may be more complicated provided that it achieves the object of even coverage at the centre.

Referring to FIG. 6*a* of the drawing there is represented therein an X-ray tube **20** from which the rays, when the tube is operating, pass through two collimators **21** and **22**. The collimator **21** is aligned with a further collimator **23** and the collimater **22** is aligned with a further collimator **24**. Between collimator **22** and **24** is located a dummy attenuator **25**. There is a gap between the collimators **21** and **23** for the location of the object to be X-rayed and in the example illustrated this gap is occupied by a plastics block **26** having a central aperture **27** for the body to be X-rayed. The plastics material may be for example that known as perspex. Two scintillators **28** and **29** are located at the ends of the collimators **23** and **24** respectively and these communicate via a light pipe **30** with a photomultiplier **31**. A chopper **33** rotatable by an electric motor **34** is arranged to allow beams to pass through the collimators **21** and **22** only alternately to produce scintillations in the scintillators **28** and **29** for detection by the photomultiplier **31**. When the apparatus is in use, the collimators **21** to **24**, the attenuator **25**, the scintillators **28** and **29**, the light pipe **30** the photo-multiplier **31**, the chopper **33** and the motor **34** are oscillated through the angle subtended by the block **26**. The X-ray source **20** does not take part in this oscillation because it produces a beam wide enough to span the block **26**. However the whole equipment is arranged to rotate slowly about the body to be examined by X radiation. Such a body is represented by the outline **32**.

The use of the scintillator **29** and the attenuator **25** provides a reference for the photo-multiplier **31**. The material of the attenuator **25** is selected to have similar absorption properties to the body **32** to be examined so that accurate transmission readings may be obtained from the X radiations which pass through this body substantially independent of the X-ray source intensity. The material in the dummy attenuator **25** compensates, to some extent for the X-ray tube spectrum drift. The space **27** between the body and block **26** is filled with a bag containing water so that the beam intensity received by the scintillator **28** is kept at constant as possible as it traverses the body **32**, thus reducing the range of the readings which the photo-multiplier **31** has to handle. The apparatus may be calibrated initially by inserting a round homogeneous body in the aperture of the block **26**. FIG. 6*b* is a similar system but the chopper is discarded and two separate detectors are used for measuring the sources and readings through the body.

The modification of FIG. 6*a* which is illustrated in FIG. 7 is intended to reduce the time required to complete an examination. According to FIG. 7 a series of photo-multipliers 31_1, 31_2 are used instead of the single photo-multiplier **31** of FIG. 6. The photo-multipliers have a common reference scintillator **29** and light pipe **30**. Each photo-multiplier has individual collimators between it and the source of X-rays **20**, the collimators being denoted by the references 21_1 and 23_1 in the case of the photo-multiplier 31_1. With this form of the invention the oscillation of the photo-multipliers and the associated collimating systems need be only a fraction of that of the apparatus shown in FIG. 6*a*. The photo multipliers could also be arranged slightly displaced downwards so that six pictures can be taken at one time. As indicated, in FIG. 8*a* the outputs of the photomultipliers 31_1, 31_2 are applied to a series of amplifiers 41_1, 41_2 . . . and thence to a serialiser **42** which feeds the plurality outputs of the amplifiers in series to an analogue-to-digital converter **43**. The digital output of the converter **43** is fed to a magnetic tape recorder **44** and thence to a digital computer **45** which is programmed to compute the absorption coefficients of the elements of a matrix notionally superimposed on the body **32** under examination. The co-efficients computed by the computer **45** are recorded by a further magnetic tape recorder **46** from which they are applied to a picture selector control device **47**. The tape produced by the computer **45** may be replaced on the tape recorder **44**, recorder **46** then being unnecessary. The output of device **47** is applied to a digital-to-analogue converter **48** and thence to a control circuit **49** which has a manual knob **50** for controlling the position of the contrast window and another manual knob **51** for controlling the width of the window. The output of the control circuit **49** is fed to a display unit **52** which includes a cathode ray tube having a screen **53**. The display unit **52** is arranged to respond to the output signals of the digital computer to build up a visual representation of the section of the object under examination. The term "window" denotes the range of signal amplitudes which is applied to the unit **52** to form the display, and the unit **52** is thus such that different absorption coefficients can be displayed on a scale from black to white. The contrast window width control knob **51** enables the full scale black to white to be occupied by a small or large critical range of absorption coefficients, and the observer may vary the position of the window by manipulation of the control knob **50**. FIGS. 8*b*, 8*c* and 8*d* illustrate the effect of varying the width and position of the window. The values of the absorption coefficients are indicated on the vertical scale in these Figures. FIG. 8*b* illustrates the case in which a wide window is used, that is to say in which the black/white range covers a wide range of values of absorption coefficients. If signals exceeding peak white are removed, for example by limiting, only tissue and tumour will show on the picture. However, as the absorption coefficient of tumour is only 10 percent greater than that of tissue both will appear as grey and it will be difficult to distinguish between them. FIG. 8*c* shows the effect of using a narrow window. In this case it is not possible to distinguish between bone and tumour but it is easy to distinguish tissue from both bone and tumour. If signals exceeding peak white are removed, only tissue will show up on the picture.

FIG. 8*d* shows the effect of altering the position of the narrow window used in FIG. 8*c*. The tumour now appears as grey while tissue exceeds peak black and bone exceeds peak white. Consequently if signals exceeding peak white and peak black are removed, only the tumour will show up in the picture. It can therefore be seen that by manipulation of the width control knob **51** and position control knob **50** the operator can eliminate from the final picture everything except the material which he wishes to examine. The display unit may also include means for displaying up to four representations of different sections at one time and provision may be made to enable the observer to dwell on one representation. A long after glow tube may be used the picture being replenished by a continuous backwards and forwards pass of the tape deck. The digital computer **45** may be an on line computer and may be remote from the magnetic tape recorders **44** and **46** being connected thereto by suitable lines or the like. Alternatively the magnetic tape recorders may be arranged to store information for computation and display at desired times.

In some cases it may be more convenient to have a direct display. This could employ a cathode ray tube store for storing the data in analogue form. Preferably, the tube should have large values of screen capacity so that the stored information may be interrogated without causing any significant change in its value. Such tubes are commonly used to provide "bright" radar displays. The summation and computing of values received from the cathode ray tube may be carried out by a simple accumulator and comparator operating a serial mode, and the output fed back to the cathode ray tube to give the necessary small additions to the change built up over the screen. A digital computer would therefore be unnecessary.

It will be appreciated that in the examples illustrated in FIGS. 6*a*, 6*b* and 7 the rays or beams of each set which are passed through the body traverse discrete paths disposed along respective radii of a sector of a circle centred at the source **20**. Moreover, as the equipment rotates, the locus of the centres of the circles (from which radiate the paths traversed by the successive sets of rays or beams) generate an arc or circle lying outside the body.

In the examples of the invention which have been described, the detecting means detect the transmission of radiation through a plurality of paths which are coplanar with the slice being examined. In some cases however some at least of the paths may be oblique to the plane of the slice and such oblique paths when it is desired used when it is desired to determine the transmission or absorption coefficients of the elements of a three dimensional matrix.

An important advantage of the invention is that the total exposure of the body is less per picture as only a narrow slit is irradiated whereas in other proposed methods an area equal to the size of the picture is exposed. In addition, the invention uses a method of decyphering the received information which is much more efficient than other methods so that more information is obtained for the same dosage.

What I claim is:

1. A method of examining at least part of the interior of a body using penetrating radiation such as X or γ rays, comprising the steps of:

a. transmitting radiation from an external source through the body in a plurality of rays traversing a plurality of respective paths to establish a first set of rays in a planar slice of said body, the cross-sectional dimensions of each ray being small in relation to the dimensions of the body;

b. transmitting radiation from said external source in further sets of rays similar to said first set of rays;

c. one of said sets being disposed in said planar slice at an initial angle or initial mean angle and the others of said sets of rays being disposed in said planar slice at angles or mean angles different from each other and from said initial angle or initial mean angle;

d. the sets of rays being such that every element of a two-dimensional matrix of elements of the body in said planar slice is intersected by a group of said rays, the group of rays being different for the different elements of the matrix, each element of said matrix having dimensions of the order of magnitude of the cross-sectional dimensions of said rays;

e. deriving from each ray emerging from the body a discrete output signal representing the sum of the transmissions or absorptions of the elements of the body intersected by the ray, thereby to derive sets of discrete output signals corresponding to the sets of rays indicative of the transmission or absorption of each element of said matrix; the number of output signals which are derived from said rays being sufficient to obtain by a process of successive approximations substantial discrimination between the transmission or absorption coefficients of neighboring elements of the matrix;

f. deriving from said output signals, by a process of successive approximations during which the approximation of the transmission or absorption coefficient of each element of the matrix is adjusted a number of times, resultant signals representing the transmissions or absorptions of the elements of said matrix, the derivation of said resultant signals including the steps of:

i. deriving an error signal responsive to a comparison of each discrete output signal and a reconstruction of said output signal predicated upon preceding approximations of the transmissions or absorptions of the elements of said matrix from which the said output signal was derived; and

ii. adjusting the respective approximations in response to said error signal;

g. producing in response to said resultant signals a representation of the transmissions or absorptions of said elements of the slice of the body.

2. A method according to claim 1 in which the steps of transmitting radiation are carried out utilizing a total number of rays substantially greater than the number of elements of said matrix.

3. A method according to claim 1 further including the steps of continuing the process of successive approximations for deriving said resultant signals until said discrete output signals have been utilized several times.

4. A method according to claim 3 further including the steps of multiplying the error signals derived during successive utilizations of said discrete output signals by a factor which is less than unity.

美国 America

3,778,614

13

5. A method according to claim 4 further including the step of reducing the factor successively for successive utilizations.

6. A method according to claim 1 in which the step of deriving resultant signals is carried out utilizing the discrete output signals taken in sets.

7. A method according to claim 1 in which the step of adjusting the approximations is carried out by distributing said error signal among the respective approximations.

8. A method according to claim 1 in which the angular separation of different sets of rays is sufficiently small that the elements of the matrix traversed by respective rays at neighboring angles in adjacent sets include a plurality of common elements.

9. A method according to claim 8 in which in deriving said resultant signals the sets of output signals are selected in such order as to correspond to sets of rays which are further apart in angle than adjacent sets of rays.

10. A method according to claim 1 in which the steps of transmitting radiation through the body are carried out such that each set of rays is disposed along respective radii of a respective sector of a circle, the locus of the centres of said circles generating a circle or arc lying outside said body.

11. A method according to claim 1 wherein step (a) is carried out by scanning said source linearly with respect to said body to establish the first set of rays traversing said body at the initial angle in the planar slice of said body, and step (b) is carried out by repeating said linear scanning of said source at a plurality of different angles around the plane, to establish the further sets of rays disposed at angles different from each other and from said initial angle.

12. A method of examining at least part of the interior of a body using penetrating radiation such as X or γ rays, comprising the steps of:

a. transmitting radiation from an external source through the body in a plurality of rays traversing a plurality of respective paths to establish a first set of rays in a planar slice of said body, the cross-sectional dimensions of each ray being small in relation to the dimensions of the body;
b. transmitting radiation from said external source in further sets of rays similar to said first set of rays;
c. one of said sets of rays being disposed in said planar slice at an initial angle or initial mean angle and the others of said sets of rays disposed in said planar slice at angles or mean angles different from each other and from said initial angle or initial mean angle;
d. the sets of rays being such that every element of a two-dimensional matrix of elements of the body in said planar slice is intersected by a group of said rays, the group of rays being different for the different elements of the matrix, each element of said matrix having dimensions of the order of magnitude of the cross-sectional dimensions of said rays;
e. deriving from each ray emerging from the body a discrete output signal representing the sum of the transmissions or absorptions of the elements of the body intersected by the ray, thereby to derive sets of discrete output signals corresponding to the sets of rays indicative of the transmission or absorption of each element of said matrix, the number of output signals which are derived from said rays being

14

sufficient to obtain by a process of successive approximations substantial discrimination between the transmission or absorption coefficients of neighboring elements of the matrix;
f. providing stored weighting signals related to different intercepts of the elements of said matrix by the rays in different sets;
g. deriving resultant signals representing the transmissions or absorptions of the elements of said matrix, the derivation of said resultant signals including the steps of successively approximating said resultant signals in response to said output signals and respective weighting signals;
h. deriving in response to the resultant signals a representation of the transmissions or absorptions of the elements of the slice of the body.

13. A method of examining at least part of the interior of a body using penetrating radiation such as X or γ rays, comprising the steps of:

a. transmitting radiation from an external source through the body in a plurality of rays traversing a plurality of respective paths to establish a first set of rays in a planar slice of said body, the cross-sectional dimensions of each ray being small in relation to the dimensions of the body;
b. transmitting radiation from said external source in further sets of rays similar to said first set of rays;
c. one of said sets of rays being disposed in said planar slice at an initial angle or initial mean angle and the others of said sets of rays being disposed in said planar slice at angles or mean angles different from each other and from said initial angle or initial mean angle;
d. the sets of rays being such that every element of a two-dimensional matrix of elements of the body in said planar slice is intersected by a group of said rays, the group of rays being different for the different elements of the matrix, each element of said matrix having dimensions of the order of magnitude of the cross-sectional dimensions of said rays;
e. deriving from each ray emerging from the body a discrete output signal representing the sum of the transmissions or absorptions of the elements of the body intersected by the ray, thereby to derive sets of discrete output signals corresponding to the sets of rays indicative of the transmission or absorption of each element in said matrix, the number of output signals which are derived from said rays being sufficient to obtain by a process of successive approximations substantial discrimination between the transmission or absorption coefficients of neighboring elements of the matrix;
f. providing stored weighting signals related to different intercepts of the elements of said matrix by the rays in different sets;
g. deriving from said discrete output signals, by a process of successive approximations, resultant signals representing the transmissions or absorptions of the elements of said matrix, the derivation of the resultant signals including the steps of:
i. producing for comparison with each output signal a reconstruction of said output signal being the sum of preceding approximations of the transmissions or absorptions of the elements of the matrix from which the said output signal was derived, multiplied by the respective weighting signals;

3,778,614

15

ii. deriving an error signal responsive to a comparison of each output signal and said reconstruction thereof;
iii. distributing said error signal among the respective approximations so as to improve said approximations;
h. providing a representation of the transmissions or absorptions of the elements of said slice of the body in response to said resultant signals.

14. A method according to claim 13 further including the step of continuing the process of successive approximations for deriving said resultant signals until the sets of discrete output signals have been utilized several times.

15. A method according to claim 14 further including the step of multiplying the error signals derived during successive utilizations of said sets of output signals by a factor which is less than unity.

16. A method of examining at least part of the interior of a body using penetrating radiation such as X or γ rays, comprising the steps of:

a. transmitting radiation from an external source through the body in a plurality of rays traversing a plurality of respective paths to establish a first set of rays in a planar slice of said body, the cross-sectional dimensions of each ray being small in relation to the dimensions of the body;
b. transmitting radiation from said external source in further sets of rays similar to said first set of rays;
c. one of said sets of rays being disposed in said planar slice at an initial angle or initial mean angle and the others of said sets of rays being disposed in said planar slice at angles or mean angles different from each other and from said initial angle or initial mean angle;
d. the sets of rays being such that every element of a two-dimensional matrix of elements of the body in said planar slice is intersected by a group of said rays, the group of rays being different for the different elements of the matrix, each element of said matrix having dimensions of the order of magnitude of the cross-sectional dimensions of said rays;
e. deriving from each ray emerging from the body a discrete output signal representing the sum of the transmissions or absorptions of the elements of the body intersected by the ray, thereby to derive sets of discrete output signals corresponding to the sets of rays indicative of the transmission or absorption of each element of said matrix, the number of output signals which are derived from said rays being sufficient to obtain by a process of successive approximations substantial discrimination between the transmission or absorption coefficients of neighboring elements of the matrix;
f. deriving in response to said discrete output signals, taken individually, by a process of successive approximations during which the approximation of the transmission or absorption coefficient of each element of the matrix is adjusted a number of times, resultant signals indicative of the transmissions or absorptions of said elements of the slice of the body; and
g. deriving in response to the resultant signals obtained from said process of successive approximations a representation of the transmissions or absorptions of the elements of said slice of the body.

16

17. A method of examining at least part of the interior of a body using penetrating radiation such as X or γ rays, comprising the steps of:

a. transmitting radiation from an external source through the body in a plurality of rays traversing a plurality of respective paths to establish a first set of rays in a planar slice of said body, the cross-sectional dimensions of each ray being small in relation to the dimensions of the body;
b. transmitting radiation from said external source in further sets of rays similar to said first set of rays;
c. one of said sets of rays being disposed in said planar slice at an initial angle or initial mean angle and the others of said sets of rays being disposed in said planar slice at angles or mean angles different from each other and from said initial angle or initial mean angle;
d. the sets of rays being such that every element of a two-dimensional matrix of elements of the body in said planar slice is intersected by a group of said rays, the group of rays being different for the different elements of the matrix, each element of said matrix having dimensions of the order of magnitude of the cross-sectional dimensions of said rays;
e. deriving from each ray emerging from the body a discrete output signal representing the sum of the transmissions or absorptions of the elements of the body intersected by the ray, thereby to derive sets of discrete output signals corresponding to the sets of rays indicative of the transmission or absorption of each element of said matrix, the number of output signals which are derived from said rays being sufficient to obtain by a process of successive approximations substantial discrimination between the transmission or absorption coefficients of neighboring elements of the matrix;
f. deriving from said output signals, by a process of successive approximations during which the approximation of the transmission or absorption coefficient of each element of the matrix is adjusted a number of times, resultant signals representing the transmissions or absorptions of the elements of said matrix, the derivation of said resultant signals including the steps of:
i. deriving an error signal responsive to a comparison of each discrete output signal and a reconstruction of said output signal predicated upon preceding approximations of the transmissions or absorptions of the elements of said matrix from which the said output signal was derived; and
ii. adjusting the respective approximations in response to said error signal, whereby a representation of the transmissions or absorptions of the elements of said slice of the body can be derived.

18. Apparatus for examining at least part of a body using radiation such as X or γ rays, comprising:

a. means including a source of radiation disposed to transmit radiation through a body to be examined, and detecting means in operative relationship with said source for detecting radiation from said source transmitted through said body in at least one ray of small cross-sectional area relative to the dimensions of the body;
b. controlling means for controlling said source and said detecting means relative to said body for transmitting through a planar slice of said body to said detecting means a plurality of rays constituting a

3,778,614

17

first set of rays and transmitting through said planar slice of said body further sets of rays similar to said first set of rays, said sets of rays being disposed in said slice at angles or mean angles different from each other;
c. said sets of rays being such that every element of a two-dimensional matrix of elements of said body in said planar slice is intersected by a group of rays, the group of rays being different for the different elements of the matrix, the elements of said matrix having dimensions of the order of magnitude of the cross-sectional dimensions of said rays;
d. circuit means coupled to said detecting means for deriving from each ray a discrete output signal representing the sum of the transmissions or absorptions of the elements of the body intersected by the respective ray, thereby to derive sets of output signals corresponding to the sets of rays indicative of the transmission or absorption of each element of said matrix, the number of output signals which are derived from said rays being sufficient to obtain by a process of successive approximations substantial discrimination between the transmission or absorption coefficients of neighboring elements of the matrix;
e. arithmetic means organized to respond to said discrete output signals and to produce, by a process of successive approximations during which the approximation of the transmission or absorption coefficient of each element of the matrix is adjusted a number of times, resultant signals representing the transmissions or absorptions of the elements of the matrix, said arithmetic means including means for deriving an error signal responsive to a comparison of each output signal and a reconstruction of said output signal predicated upon preceding approximations of the transmissions or absorptions of the elements of the matrix from whic the said output signal was derived and for adjusting the respective approximations in response to said error signal; and
f. means responsive to said resultant signals for producing a representation of the transmissions or absorptions of said elements of said slice of the body.

19. Apparatus according to claim 18, wherein said controlling means controls said source and said detecting means relative to said body to transmit the sets of rays in said planar slice such that the rays of each set are disposed along respective radii of a respective sector of a circle, the locus of the centres of said circles generating a circle or arc lying outside said body.

20. Apparatus according to claim 18, wherein said arithmetic means includes:

i. means for storing preceding approximations of the transmissions or absorptions of the elements of said matrix;
ii. means for deriving for each discrete output signal a reconstruction of said output signal predicated upon preceding approximations;
iii. means for deriving an error signal responsive to a comparison of each discrete output signal and said reconstruction thereof;
iv. means for adjusting the respective approximations stored in said storing means in response to said error signals.

21. Apparatus according to claim 18 further including storage means for storing said discrete output signals.

18

22. Apparatus according to claim 21 wherein said arithmetic means further includes means for selecting said stored discrete output signals in sets from said storage means.

23. Apparatus according to claim 22 wherein said controlling means is arranged to cause the angular separation of the different sets of rays to be sufficiently small that the elements of the matrix traversed by respective rays at neighboring angles in adjacent sets include a plurality of common elements and wherein said means for selecting said output signals in sets is arranged to select the sets in such order as to correspond with sets of rays which are further apart in angle than adjacent sets of rays.

24. Apparatus according to claim 18 further comprising a store of weighting signals related to the intercepts of the elements of the matrix by the rays in different sets determined by said controlling means and wherein said means for deriving an error signal comprises means for summing the preceding approximations of the transmissions or absorptions of the elements of the matrix from which the respective output signal was derived, multiplied by the respective weighting signals.

25. Apparatus according to claim 18 wherein controlling means for controlling said source and said detecting means linearly scans said body to establish said first set of rays and repeats said linear scanning of said source and said detecting means at a plurality of different angles round said plane to establish said further sets of rays.

26. Apparatus for examining at least part of a body using radiation such as X or γ rays, comprising:

a. means including a source of radiation disposed to transmit radiation through a body to be examined, and detecting means in operative relationship with said source for detecting radiation from said source transmitted through said body in at least one ray of small cross-sectional area relative to the dimensions of the body;
b. controlling means for controlling said source and said detecting means relative to said body for transmitting through a planar slice of said body to said detecting means a plurality of rays constituting a first set of rays and for transmitting through said planar slice of said body to said detecting means further sets of rays similar to said first set of rays, said sets of rays being disposed in said slice at angles or mean angles different from each other;
c. said sets of rays being such that every element of a two-dimensional matrix of elements of said body in said planar slice is intersected by a group of rays, the group of rays being different for the different elements of the matrix, the elements of said matrix having dimensions of the order of magnitude of the cross-sectional dimensions of said rays;
d. circuit means coupled to said detecting means for deriving from each ray a discrete output signal representing the sum of the transmissions or absorptions of the elements of the body intersected by the respective ray, thereby to derive sets of output signals corresponding to the sets of rays indicative of the transmission or absorption of each element of said matrix;
e. a store for weighting signals related to the intercepts of the elements of said matrix by the rays in different sets determined by said controlling means;

3,778,614

19

f. arithmetic means for deriving resultant signals representing the transmissions or absorptions of the elements of said matrix, said arithmetic means being organized to derive said resultant signals by a process of successive approximations in response to said output signals and said weighting signals;
g. means responsive to said resultant signals for producing a representation of the transmissions or absorptions of said elements of said slice of said body.

27. Apparatus for examining at least part of a body using radiation such as X or γ rays, comprising:

a. means including a source of radiation disposed to transmit radiation through a body to be examined, and detecting means in operative relationship with said source for detecting radiation from said source transmitted through said body in at least one ray of small cross-sectional area relative to the dimensions of the body;
b. controlling means for controlling said source and said detecting means relative to said body for transmitting through a planar slice of said body to said detecting means a plurality of rays constituting a first set of rays and for transmitting through said planar slice of said body further sets of rays similar to said first set of rays, said sets of rays being disposed in said slice at angles or mean angles different from each other;
c. said sets of rays being such that every element of a two-dimensional matrix of elements of said body in said planar slice is intersected by a group of rays, the group of rays being different for the different elements of the matrix, the elements of said matrix having dimensions of the order of magnitude of the cross-sectional dimensions of said rays;
d. circuit means coupled to said detecting means for deriving from each ray a discrete output signal representing the sum of the transmissions or absorptions of the elements of the body intersected by the respective ray, thereby to derive from the sets of rays corresponding sets of output signals indicative of the transmissions or absorptions of each element of the matrix, the number of output signals which are derived from said rays being sufficient to obtain by a process of successive approximations substantial discrimination between the transmission or absorption coefficients of neighboring elements of the matrix;
e. arithmetic means for deriving resulting signals indicative of the transmissions or absorptions of the elements of the matrix in response to said discrete output signals, taken individually, by a process of successive approximations during which the approximation of the transmission or absorption coefficient of each element of the matrix is adjusted a number of times;
f. means responsive to resultant signals obtained from said process of successive approximations for producing a representation of the transmissions or ab-

20

sorptions of the elements of said slice of the body.

28. Apparatus for examining at least part of a body using radiation such as X or γ rays, comprising:

a. means including a source of radiation disposed to transmit radiation through a body to be examined, and detecting means in operative relationship with said source for detecting radiation from said source transmitted through said body in at least one ray of small cross-sectional area relative to the dimensions of the body;
b. controlling means for controlling said source and said detecting means relative to said body for transmitting through a planar slice of said body to said detecting means a plurality of rays constituting a first set of rays and for transmitting through said planar slice of said body further sets of rays similar to said first set of rays, said sets of rays being disposed in said slice at angles or mean angles different from each other;
c. said sets of rays being such that every element of a two-dimensional matrix of elements of said body in said planar slice is intersected by a group of rays, the group of rays being different for the different elements of the matrix, the elements of said matrix having dimensions of the order of magnitude of the cross-sectional dimensions of said rays;
d. circuit means coupled to said detecting means for deriving from each ray a discrete output signal representing the sum of the transmissions or absorptions of the elements of the body intersected by the respective ray, thereby to derive from the sets of rays corresponding sets of discrete output signals indicative of the transmission or absorption of each element of said matrix, the number of output signals which are derived from said rays being sufficient to obtain by a process of successive approximations substantial discrimination between the transmission or absorption coefficients of neighboring elements of the matrix;
e. arithmetic means organized to respond to said discrete output signals and to produce by a process of successive approximations during which the approximation of the transmission or absorption coefficient of each element of the matrix is adjusted a number of times, resultant signals representing the transmissions or absorptions of the elements of the matrix, said arithmetic means including means for deriving an error signal responsive to a comparison of each output signal and a reconstruction of said output signal predicated upon preceding approximations of the transmissions or absorptions of the elements of the matrix from which the said output signal was derived and for adjusting the respective approximations in response to said error signal, whereby a representation of the transmissions or absorptions of the elements of said slice of the body can be derived.

* * * * *

RÉPUBLIQUE FRANÇAISE
INSTITUT NATIONAL
DE LA PROPRIÉTÉ INDUSTRIELLE
PARIS
A1
N° de publication
(A n'utiliser que pour les commandes de reproduction).
2 266 222
DEMANDE
DE BREVET D'INVENTION
N° 74 10191

智能卡的发明

invention

Smart card

罗兰·莫瑞诺

Roland Moreno

法国人罗兰·莫瑞诺（Roland Moreno）是一名科学记者，1974 年，在他 29 岁，还是一名记者的时候，发明了智能卡，被誉为“智能卡之父”。

智能卡是一种用于身份识别且具有高度安全性的集成电路卡，它是将具有存储控制及数据处理功能的集成电路芯卡镶嵌入塑料卡内形成的。

智能卡已在全球范围被广泛应用，除了用于银行卡外，已被广泛用于各行各业。

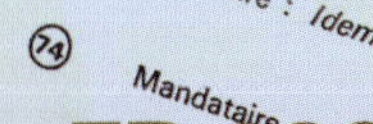

FR 2266222

第十九篇 智能卡

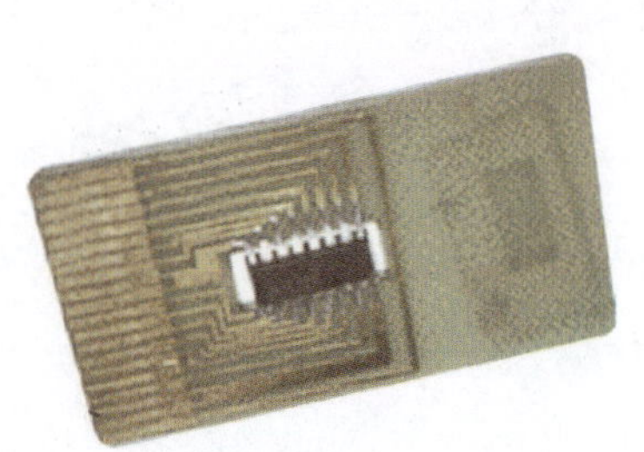

智能卡，即通常所说的IC卡（Integrated Circuit Card），是指将集成电路芯片固封在塑料基片中的卡片，是一种功能多样、用途广泛的电子卡片。它的外形和尺寸同普通名片差不多，小巧玲珑，携带方便、使用简捷。IC卡的基片是由聚氯乙烯硬质塑料制成的，内装集成电路芯片。因集成电路的英文缩写为IC，所以称为IC卡。它可与多种终端设备连接使用，具有多种功能。

1969年12月，日本的有村国孝（Kunitaka Arimura）提出一种制造安全可靠的信用卡的方法，并于1970年申请了专利，称为ID卡（Identification Card），可被看作是IC卡的最初设想。而智能卡究竟是谁最先发明的，至今仍争论不休。其中一种观点认为最先发明者是德国电气工程师尤尔根·戴德罗夫（Jurgen Dethlof）和赫尔穆特·格罗特罗普（Helmut Groettrup），他们在1969年就向德国专利商标局提交了名称为“带有集成电路卡片”的专利申请。1977年，戴德罗夫又进一步提交了名为“带有微处理功能的卡片”的专利申请，并于1993年获得德国DE2760486号专利权，如图19-1、图19-2所示。

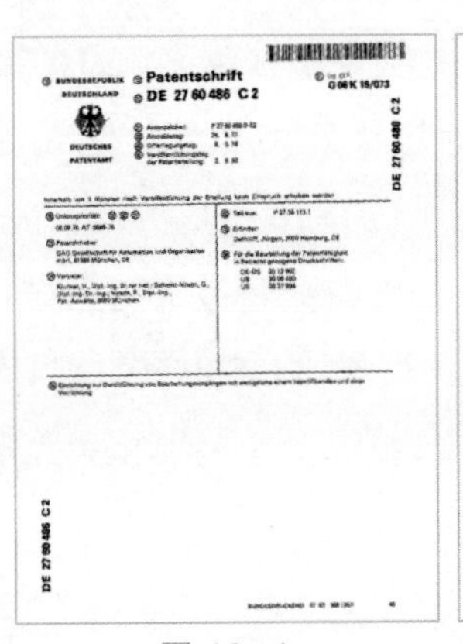

Patentschrift
DE 27 60 486 C2

图 19-1

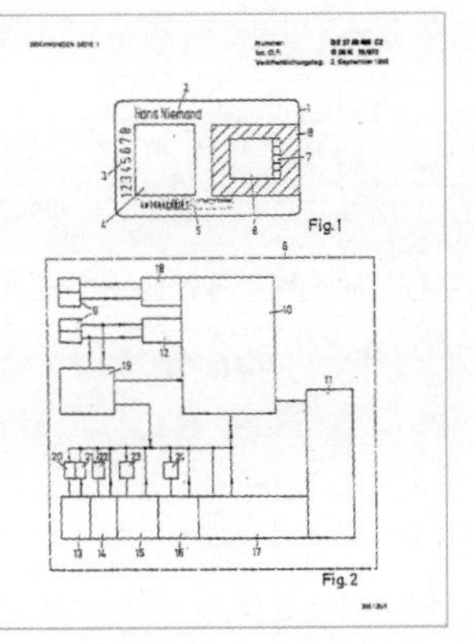

图 19-2

但是大多数公众认为，1974 年法国的罗兰 · 莫瑞诺（Roland Moreno，1945-2012 年）发明的带集成电路芯片的塑料卡片是智能卡的最先发明。

罗兰 · 莫瑞诺曾是一名科学记者，当时人们使用的银行磁卡易消磁，存储容量小，存储信息易被篡改，安全性和可靠性都比较差。罗兰考虑到当时的信息技术，特别是微电子技术的集成电路技术的发展，发明了用集成电路来存储信息的新方法，大大提升了信息存储的安全性。1974 年 3 月 25 日，他向法国提交了名为“电子控制方法与装置”的专利申请，并于 1975 年 10 月 24 日获得法国 FR2266222 号专利的公布，如图 19-3 至 19-5 所示。

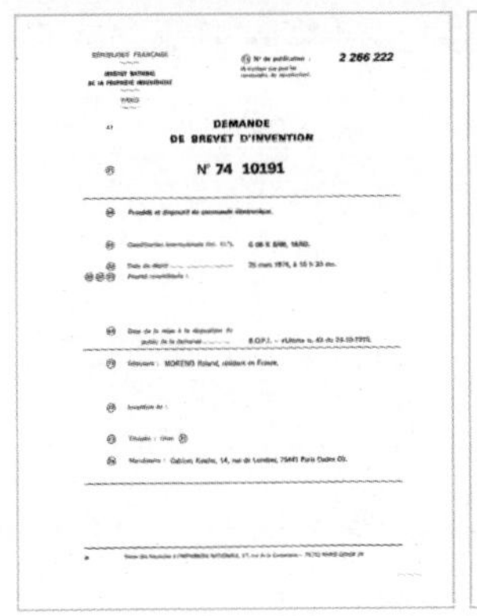

2 266 222
DEMANDE
DE BREVET D'INVENTION
N° 74 10191

图 19-3

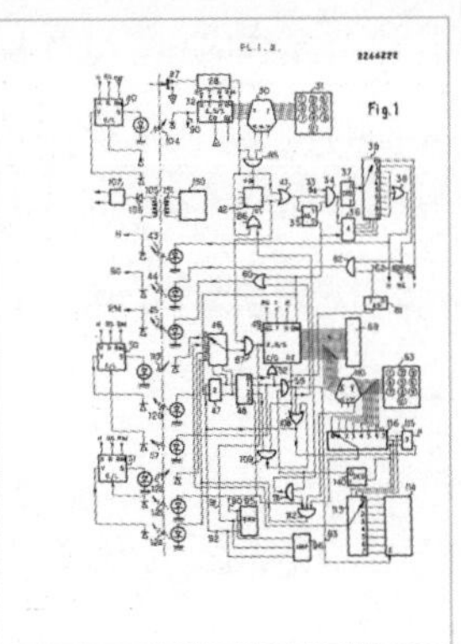

图 19-4

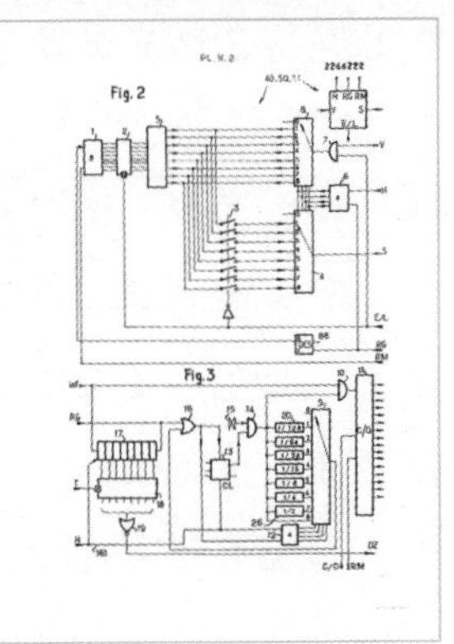

图 19-5

罗兰·莫瑞诺非常重视保护自己的创新成果，先后就这一发明在多个国家提出专利申请，并充分利用优先申请期（申请之日起12个月）修正申请案，在有效期最后两天的1975年3月21日，他以三件法国优先申请为基础向美国提交名为“数据存储方法与数据存储系统”的专利申请，并于1976年7月27日获得美国US3971916号专利权。而最早在法国申请的“电子控制方法与装置”（FR2266222号）却是在1980年才获得法国的专利授权。

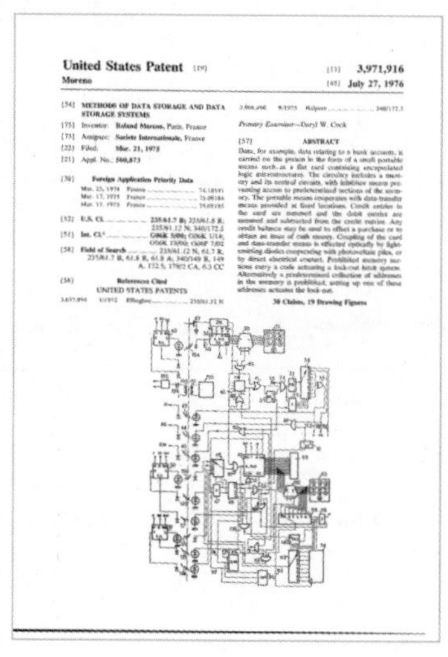

United States Patent [19]
Moreno
[11] 3,971,916
[45] July 27, 1976

[54] METHODS OF DATA STORAGE AND DATA STORAGE SYSTEMS
[75] Inventor: Roland Moreno, Paris, France
[73] Assignee: Societe Internationale, France
[22] Filed: Mar. 21, 1975
[21] Appl. No.: 560,873
[30] Foreign Application Priority Data
[52] U.S. Cl.
[51] Int. Cl.
[58] Field of Search
[56] References Cited
UNITED STATES PATENTS

Primary Examiner—Daryl W. Cook

[57] ABSTRACT

38 Claims, 19 Drawing Figures

图 19-6

从美国 US3971916 号专利说明书中可以看出，如图 19-6 所示，罗兰·莫瑞诺详细说明了磁条卡的缺陷及存储内容可能由于热和磁场而消磁或受损；存储容量相对比较小，身份识别数据位数有限；存储的内容会被一个相对简单的装置，如塑料透镜，磁针进行修改；磁条要与所放置的读写设备的阅读写头靠近并相对移动，可靠性受到一定影响。他在申请中针对性地提出了解决方案，一种新的存储数据的方法和数据存储系统。该系统包括便携式电子装置和数据传输装置，数据传输装置与中心计算机相连，数据传输装置包括操作时将计算机与便携式的电子装置进行数据传输的传输器。该专利申请还进一步提出，所述的便携式电子装置可以是卡状、环状、管状或其他形状的。而作

为典型例子提出是卡片式的，就是封装有逻辑微结构集成电路的塑料卡片，而且其所揭示的电子微结构卡，不再是单纯的存储卡，而是具有存储、处理、控制、安全于一体的智能型卡。

莫瑞诺最初是想制作一种带有芯片和徽章、可以读取和传输信息的贵族图章戒指，但事实证明在当时这一想法很不现实。他很快就简化了设计，将微芯片镶嵌在塑料卡片中。1976 年，莫瑞诺在装配玩具上安装了设备，第一次向世人证明了智能卡可以在金融交易中使用。由于首次启动成本高，他发明的智能卡初期几乎没有什么人使用，直到 1983 年法国电信开始在公共电话领域大规模使用智能卡，才使得智能卡在法国得到广泛使用。

为了推广智能卡技术，莫瑞诺在 1974 年成立了 Innovatron 高科技公司来实施、推广这项专利技术。自 1992 年以来，Innovatron 高科技公司和主要运输网络 RATP 和 SNCF 合作研发，拓展了智能卡技术的应用领域，开发了适用于公共交通上使用的非接触式通信接口技术。在项目研发过程中，Innovatron 高科技公司申请了多项专利保护自身的研发成果，成果应用在欧洲远程售票项目 Icare 和 Calypso 上。这一时期比较有代表性的专利是莫瑞诺 1992 年在法国的申请，专利名称为“终端与便携式组件间远程数据交换系统”，于 1997 年获得法国 FR2689997 号专利权，如图 19-7 所示。该专利涉及一种通常是可移动的便携式组件与一个或一组通常是固定的终端之间进行远程数据交换的系统。卡和可读卡的便携式设备是分开的，改进之处在于可实现用智能卡远程访问高速公路和公共交通的付费终端，完成付费，即现在通常说的刷卡付费。

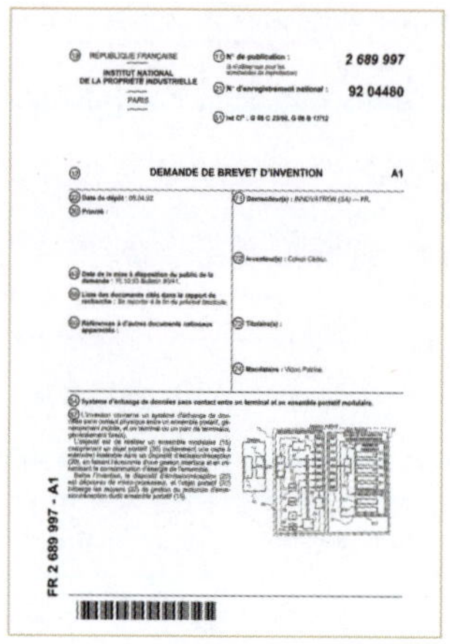

RÉPUBLIQUE FRANÇAISE
INSTITUT NATIONAL DE LA PROPRIÉTÉ INDUSTRIELLE
PARIS

N° de publication : 2 689 997
N° d'enregistrement national : 92 04480

DEMANDE DE BREVET D'INVENTION A1

Date de dépôt : 09.04.92
Priorité :
Demandeur(s) : INNOVATRON (SA) — FR.
Date de la mise à disposition du public de la demande :
Liste des documents cités dans le rapport de recherche :
Références à d'autres documents nationaux apparentés :
Inventeur(s) :
Titulaire(s) :
Mandataire :

Système d'échange de données sans contact entre un terminal et un ensemble portatif modulaire.

FR 2 689 997 - A1

图 19-7

Innovatron 高科技公司还不失时机地将自家专利技术转化为技术标准，实现专利技术的国际标准化。我们从 Innovatron 高科技公司网站上（www.innovatron.fr）可以详细了解到该公司与 ISO14443B 标准[①]有关信息及其提供对外许可的专利技术清单，参见表 19-1。

表 19-1 Innovatron 高科技公司提供对外许可的专利技术

序号	美国专利号	欧洲专利号	技术内容
BID119	US6636146	EP0901670	终端与远程赋能 IC 卡间通过非接触通信交换数据的系统
BID120	US6646543	EP1016023	在非接触数据交换系统中处理冲突的方法
BID1	US6690263	EP0976084	非接触式数据交换系统的改进的冲突管理方法

我们从专利文献中发现，Innovatron 高科技公司这 3 项技术在 1996 ~ 1998 年，首先向法国提交了 3 件专利申请，随后通过欧洲专利局和国际专利合作条约（PCT）两个国际组织进行了多个国家的专利申请，并获得多国专利授权。2003 ~ 2004 年，我国也向这 3 件专利授予了专利权，如图 19-8、图 19-9、图 19-10 所示。

① ISO14443B是（非接触式IC标准）协议。

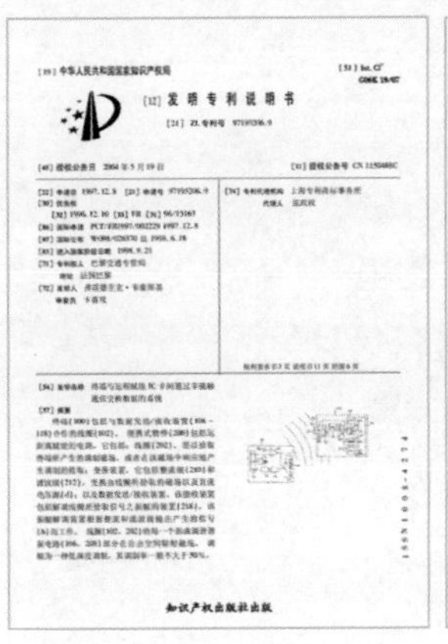

[19] 中华人民共和国国家知识产权局

[12] 发明专利说明书

知识产权出版社出版

图 19-8

[19] 中华人民共和国国家知识产权局

[12] 发明专利说明书

知识产权出版社出版

图 19-9

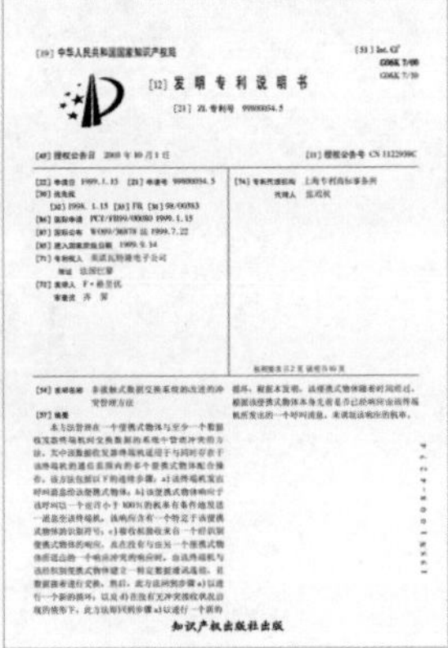

[19] 中华人民共和国国家知识产权局

[12] 发明专利说明书

知识产权出版社出版

图 19-10

Innovatron 高科技公司拥有的这些专利技术，成为非接触式智能卡生产中必须采用的技术，各生产商都需要获得 Innovatron 高科技公司的专利许可。从该公司网站上列出的专利许可生产商的名单中，可以看到夏普、三星这样的知名跨国公司，来自中国的握奇、复旦微电子等多家单位也使用了该公司的专利技术。

随着技术的发展，智能卡正由“一卡专用”发展为“一卡多用”，逐渐被广泛应用在金融、通信、交通、安全、证照等领域。现今，我们日常生活中处处离不开智能卡，公交卡、银行卡、信用卡和医疗卡这些承载着重要信息的卡片总会出现在人们的钱包中。因为智能卡的发明，法国人视莫瑞诺为国家英雄，2009 年莫瑞诺获法国荣誉军团勋章。

1975. FR 2266222

——电子控制方法与装置

法国

France

法国

France

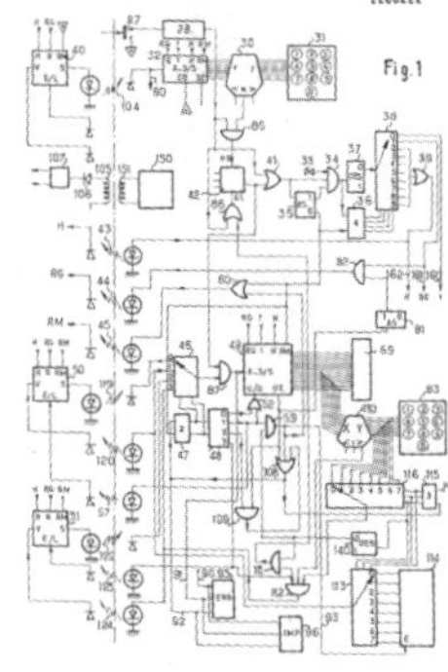

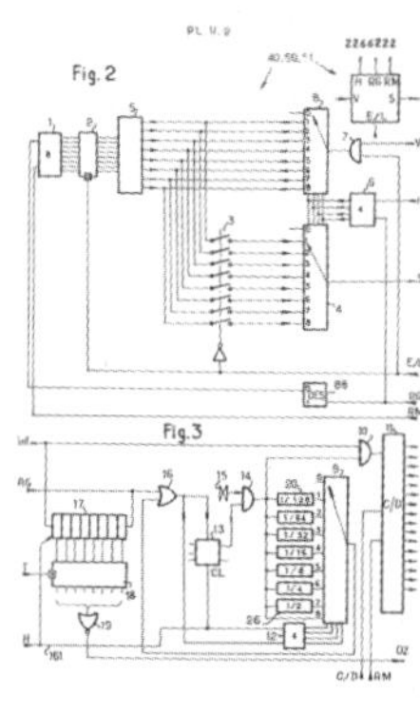

Raymond Damadian
3,789,832
Feb. 5, 1974
磁共振成像
雷蒙德·达马蒂安
Raymond Damadian
雷蒙德·达马蒂安 1936 年出生于美国纽约，1956 年获得数学专业的理学士学位，接着转向医学并在 1960 年获得爱因斯坦医学院的医学博士学位。后进入美国纽约下城医学院，在这里，他对活体细胞中钠和钾的研究，引导他开始第一个用到磁共振试验，并最终在 1977 年制成第一台磁共振人体扫描仪，为世界科学和医学作出了重大贡献。
US 3789832
POWER AMPLIFIER
RECEIVER
PHASE DETECTOR
LOW PASS
VARIABLE VOLTAGE SUPPLY
ELECTROMAGNET
VERTICAL INPUT
OSCILLISCOPE
SYNC
www.sipo.gov.cn

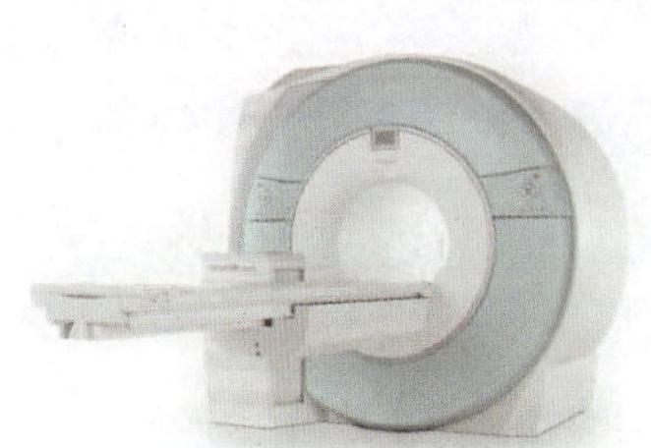

第二十篇 磁共振成像

在 2003 年 10 月 10 日的《纽约时报》和《华盛顿邮报》上，同时出现了佛纳公司的一则整版广告："雷蒙德·达马蒂安（Raymond Damadian，又译达马迪安），应当与彼得·曼斯菲尔德（Peter Mansfield）和保罗·劳特布尔（Paul Lauterbur）分享 2003 年诺贝尔生理学或医学奖。没有他，就没有磁共振成像技术。"该广告因指责诺贝尔奖委员会"篡改历史"而引起广泛争议。

雷蒙德·达马蒂安（1936– ），出生于美国纽约。他 1956 年获得数学专业的理学士，接着转向医学并在 1960 年获得爱因斯坦医学院的医学博士。后来进入美国纽约下城医学院，在这里他对活体细胞中钠和钾的研究引导他开始第一个用到磁共振的试验。1970 年，他发现在磁激条件下，磁共振对健康细胞和癌症细胞原子核的影响不同，它们从高能状态返回平衡状态时，癌细胞原子核所用的时间要长得多。1972 年，达马蒂安首先在美国提出了名为"用于癌组织检测的设备和方法"的专利申请。

达马蒂安发明的就是磁共振成像设备。磁共振成像是医学影像技术的一种，它的出现和发展是医学领域的重大突破，为临床诊断、治疗和医学研究提供了重要的资料和依据。与普通 X 射线或超声波成像相比，磁共振成像的最大优点是对人体没有任何伤害，是一种安全、快速、准确的临床诊断方法。

追溯磁共振成像技术的起源，还要从1946年美国斯坦福大学物理学家布罗克（Bloch F.）和哈佛大学物理学家伯塞尔（Purcell E .M. ）通过实验发现磁共振现象说起。1971年，在美国纽约州立大学任教的达马蒂安首次提出用磁共振波谱仪检查人体正常组织和癌变组织，并在《科学》杂志上发表了题为“磁共振信号可检测疾病”和“癌组织中氢的TI时间延长”的论文，为磁共振技术在医学领域的应用揭开了新的篇章，也奠定了自己在磁共振成像技术发展史上的奠基人地位。

达马蒂安通过专利申请，记录了自己在磁共振成像技术上的成就。1972年他首先在美国提出专利申请，并于1974年获得名为“探测人体组织中癌的设备和方法”的美国US3789832号专利权，如图20-1、图20-2所示。他看到这项技术具有广阔的商业前景，用现在的话说就是他实施了有效的专利布局，充分利用美国优先申请日，在还有最后两天的1973年3月15日向加拿大提出专利申请，并于1977年1月25日获得加拿大CA1004297号专利权，凭借这一技术成功占领了北美市场。他在1977年制成第一台磁共振人体扫描仪，命名为“Indomitable”（中文含义为“不屈服的”），完成了第一幅人体全身磁共振成像，产生了一种革命性的医学诊断工具。

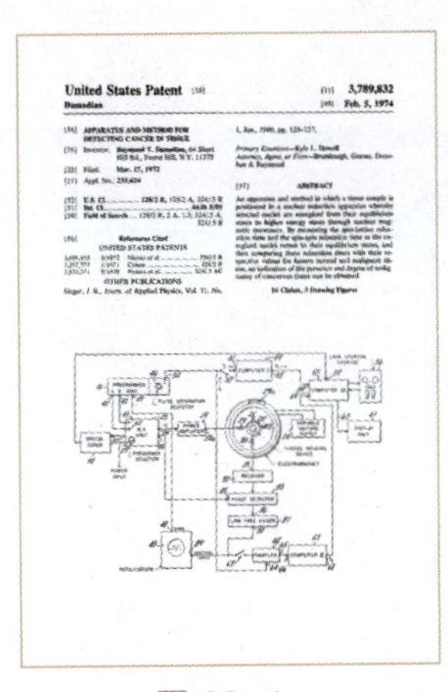

United States Patent [19]
Damadian
[11] 3,789,832
[45] Feb. 5, 1974

图20-1

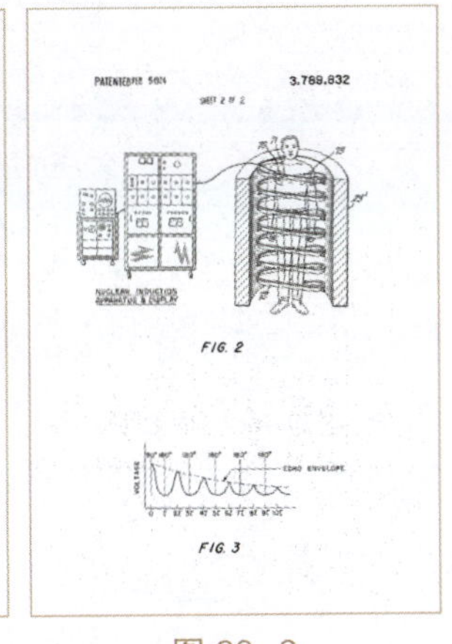

图20-2

从这篇专利文献中，我们可以发现是达马蒂安首次公开了磁共振

成像系统，具体包括：磁体系统、梯度磁场系统、射频系统和计算机系统。利用人体中遍布全身的氢原子在外加的强磁场内受到射频脉冲的激发，产生磁共振现象，经过空间编码技术，用探测器检测并接受以电磁形式放出的磁共振信号，输入计算机，经过数据处理转换，最后将人体各组织的形态形成图像，以作诊断。

作为“磁共振成像技术”诺贝尔奖获得者之一的美国科学家保罗·劳特布尔（Paul Lauterbur），是在1973年提出利用磁场和射频相结合方法获得显微镜磁共振图像技术的设想的，并利用此技术获得了二维磁共振图像。但在专利文献中，劳特布尔关于“磁共振成像技术”最早的专利申请是1990年向美国提出的名为“一种计算空间编码光谱信号的局部磁共振谱的计算方法”的美国US5081992号专利，如图20-3所示。

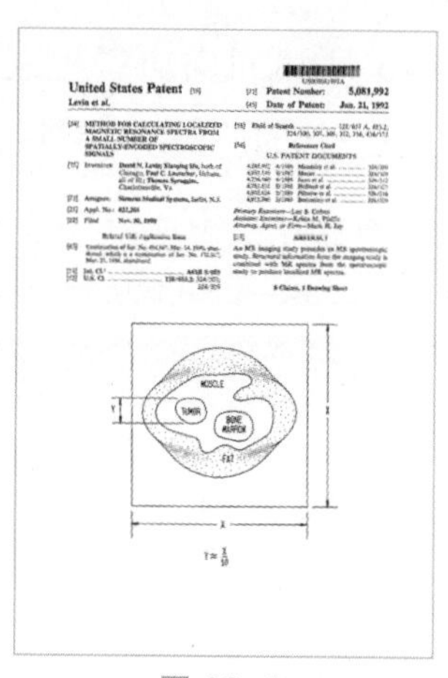

United States Patent [19]
Levin et al.
[11] Patent Number: 5,081,992
[45] Date of Patent: Jan. 21, 1992

[54] METHOD FOR CALCULATING LOCALIZED MAGNETIC RESONANCE SPECTRA FROM A SMALL NUMBER OF SPATIALLY-ENCODED SPECTROSCOPIC SIGNALS

[56] References Cited
U.S. PATENT DOCUMENTS

图 20-3

“磁共振成像技术”另一位诺贝尔奖获得者英国科学家彼得·曼斯菲尔德（Peter Mansfield），是于1974年在进一步验证和改进劳特布尔的方法基础上，发现不均匀磁场的快速变化可以使上述方法更快地绘制成物体内部结构图像，并证明了可以用显微镜数学方法分析获得数据，为利用计算机快速绘制图像奠定了基础。他虽与达马蒂安同在20世纪70年代申请了两件关于磁共振成像技术方面的专利，但比达马蒂安晚了两年。一件是1974年申请的名为“使用磁共振成像”

的美国 US4021726 号专利，另一件是 1976 年申请的名为“磁共振设备与方法”的美国 US4115730 号专利，而这件专利的在先技术正是达马蒂安的 US3789832 号专利，如图 20-4 所示。

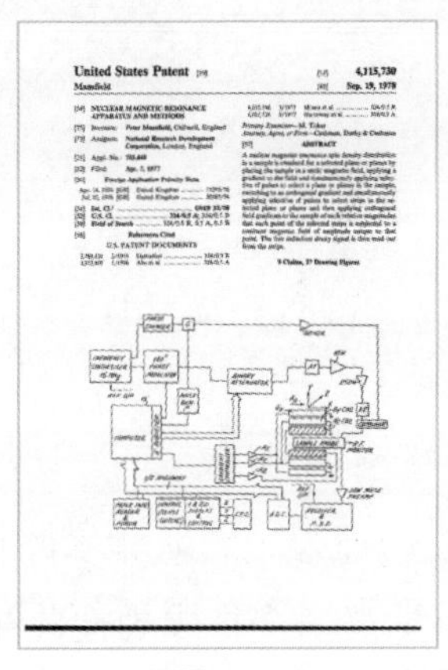

United States Patent [19] [11] 4,115,730
Mansfield [45] Sep. 19, 1978

[54] NUCLEAR MAGNETIC RESONANCE APPARATUS AND METHODS

[57] ABSTRACT

U.S. PATENT DOCUMENTS

图 20-4

历史上，对磁共振成像技术发明权的归属问题已争论了许多年，而且争得颇为激烈。达马蒂安为了维护自己作为磁共振发明人的地位，不惜大兴诉讼，曾经控告生产磁共振扫描仪的通用电气公司侵犯其专利权。官司从 1992 年打起，直到 1997 年才最终由美国联邦最高法院判决达马蒂安胜诉，通用电气公司因侵犯其美国 US3789832 号专利，而向达马蒂安的 FONAR 公司赔付巨额赔偿 1.15 亿美元。

达马蒂安在磁共振领域的研究和创造一直没有停止，1978 年他就进一步改进的成果在美国以“磁共振扫描和成像设备和方法”为名再次申请专利，并于 1982 年获得美国 US4354499 号专利权，如图 20-5 所示。1978 年他创立了 FONAR 公司，使他的发明市场化，生产磁共振扫描设备，并于 20 世纪 80 年代初获得美国国家食品管理局批准。

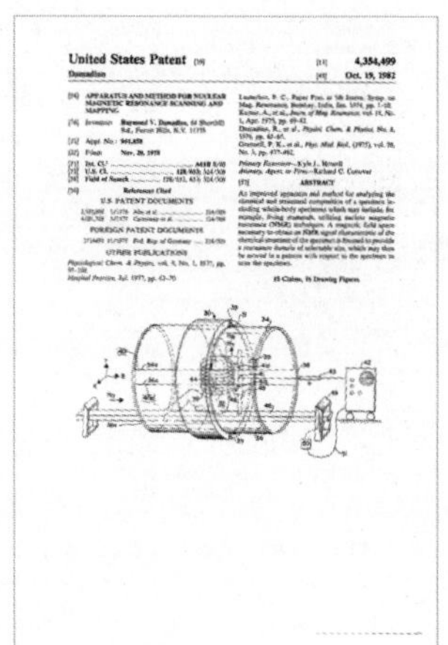

United States Patent [19]

Damadian

[11] 4,354,499

[45] Oct. 19, 1982

[54] APPARATUS AND METHOD FOR NUCLEAR MAGNETIC RESONANCE SCANNING AND MAPPING

References Cited

U.S. PATENT DOCUMENTS

FOREIGN PATENT DOCUMENTS

OTHER PUBLICATIONS

[57] ABSTRACT

图 20-5

或许，达马蒂安没能与彼得·曼斯菲尔德和保罗·劳特布尔一起分享 2003 年的诺贝尔医学奖，这将成为他此生的一个遗憾。然而，他为人类社会带来的巨大影响是伟大的，是有目共睹的。

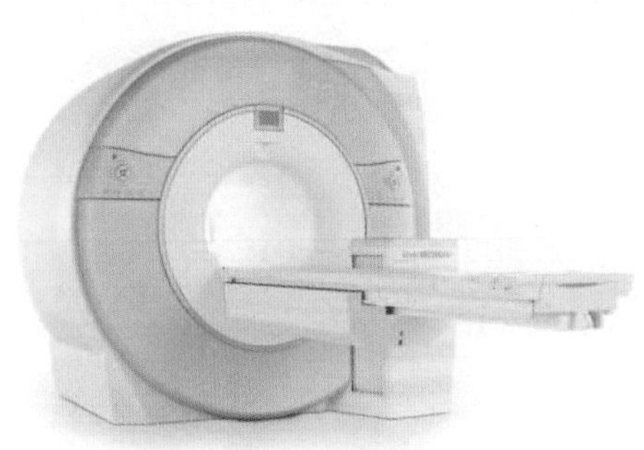

1975.US 3789832

——探测人体组织中癌的设备和方法

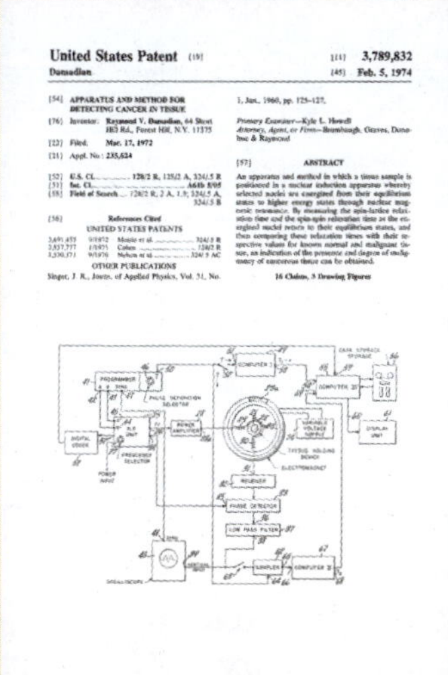

United States Patent [19] [11] **3,789,832**

Damadian [45] **Feb. 5, 1974**

[54] APPARATUS AND METHOD FOR DETECTING CANCER IN TISSUE

[76] Inventor: Raymond V. Damadian, 64 Short Hill Rd., Forest Hill, N.Y. 11375

[22] Filed: Mar. 17, 1972

[21] Appl. No.: 235,624

[52] U.S. Cl. .... 128/2 R, 125/2 A, 324/.5 R

[51] Int. Cl. A61b 5/05

[58] Field of Search ... 128/2 R, 2 A, 1.5; 324/.5 A, 324/.5 B

[56] References Cited

UNITED STATES PATENTS

3,691,453	[illegible]	[illegible] et al.	324/.5 R
3,557,777	[illegible]	Cohen	128/2 R
3,530,371	[illegible]	Nelson et al.	324/.5 AC

OTHER PUBLICATIONS

Singer, J. R., Journ. of Applied Physics, Vol. 31, No. 1, Jan., 1960, pp. 125–127.

Primary Examiner—Kyle L. Howell
Attorney, Agent, or Firm—Brumbaugh, Graves, Donohue & Raymond

[57] ABSTRACT

An apparatus and method in which a tissue sample is positioned in a nuclear induction apparatus whereby selected nuclei are energized from their equilibrium states to higher energy states through nuclear magnetic resonance. By measuring the spin-lattice relaxation time and the spin-spin relaxation time as the energized nuclei return to their equilibrium states, and then comparing these relaxation times with their respective values for known normal and malignant tissue, an indication of the presence and degree of malignancy of cancerous tissue can be obtained.

16 Claims, 3 Drawing Figures

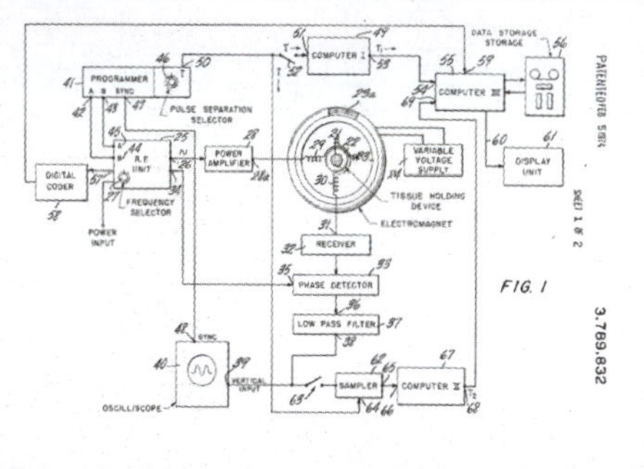

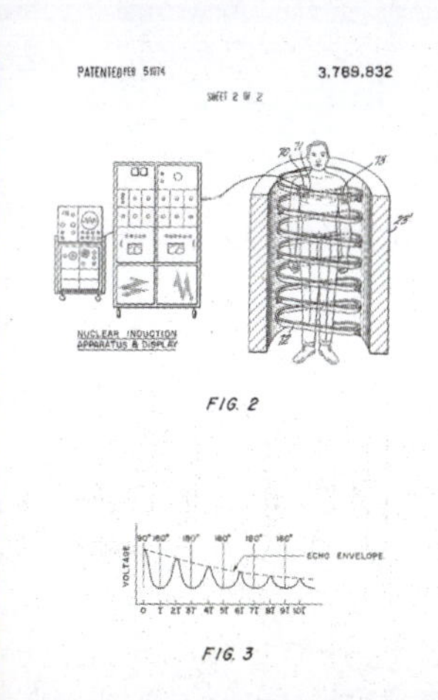

美国

America

数码相机的发明 Digital camera invention

赛尚 Steven J. Sasson

世界上第一台数码相机是由柯达应用电子研究中心，Steven J Sasson于1975年发明的，原型机名称为“手持电子照相机”。这台最古老的相机曝光时间为50毫秒，记录一张影像需要23秒，每盒磁带可存储30张照片。

相机通过拥有10000像素（按100x100的阵列排列）的CCD拍摄影像，每个像素占4个位——由0和1组成的四位数组合，表示照片中的每一个点。一旦拍摄完毕，影像便会经过数字化处理并存储到相机中的内存缓冲区。从这里，照片便可记录到更具永久性的存储器内，以便从相机上取下来进行播放。

US 4131919

www.sipo.gov.cn

第二十一篇 数码相机

法国艺术家达盖尔（Daguerre）发明世界上第一台相机距今已经有 170 多年了，而数码相机的出现只有短短几十年，但其技术发展速度颠覆了传统的相机，给我们的生活带来了巨大的改变。

数码相机无须胶卷就可以进行拍摄，相机内最重要的元件是一种光感应式的电荷耦合器件（Charge-coupled Device，CCD），获取的图像通常会先储存在数码存储设备中。1969 年，贝尔实验室的维拉·波义耳（Willard S. Boyle）和乔治·史密斯（George E. Smith）将影像电话和半导体气泡式存储技术结合，设计了能沿着一片半导体表面传导电荷的“电荷气泡器”，随即发现光电效应能使此种元件表面产生电荷而组成数位影像。到了 20 世纪 70 年代，贝尔实验室的研究人员应用简单的线性装置捕捉影像，发明了 CCD。当时发明 CCD 的目的是改进存储技术，元件本身也被当作单纯的存储器使用。随后人们认识到，CCD 可以利用光电效应来拍摄并存储图像。CCD 的诞生促进了数码相机的发明。

第一台数码相机原型机是伊士曼柯达公司（Eastman Kodak Company，简称柯达公司）的工程师赛尚（Steven J. Sasson，1950- ）发明的。当时的柯达公司是世界上最大的影像产品及相关服务的生产和供应商，总部位于美国纽约州罗切斯特市。20 世纪 70 年代，柯达公司在传统胶片相机领域已经发展到一个非常成熟的阶段，因此开始

尝试数码相机的开发项目，设想不通过胶卷也能拍摄影像。1973 年，赛尚从仑士勒理工大学电子工程系硕士毕业后加入柯达，成为一名柯达应用电子研究中心的工程师。1974 年，赛尚负责研发“电子静态照相机”项目。这个项目对于当时的柯达公司来说是一个比较小的项目，颇具有实验性，资金和人员投入都比较少，只有塞尚和另外两位技术工程师参与。

在当时开发数码相机项目存在着很多困难，CCD 很难控制，A/D 转换器也很难制造，数码存储介质难于获取，而且容量很小。由于当时还未进入个人计算机时代，显示设备也需要量身定做。正是这些技术上的难题让赛尚及其同事用了一年时间才拼装完成可数码成像的照相机。这台数码相机的 CCD 阵列只有 1 万像素，成像非常粗糙。此外，它和现在常见的数码相机相比笨重得多，它长 20.9 厘米、宽 15.2 厘米、高 22.5 厘米，重 3.9 千克，拍摄的时候非常耗电，需要 16 节 AA 电池供应电力。这台数码相机的问世标志着世界上第一部无胶卷手持相机的诞生。通过电子方式拍摄黑白静像，并将它们记录到盒式磁带上，磁带可以从相机内取下并插入播放设备，从而能够在电视上观看。赛尚使用这台相机获得了有史以来第一张数字化影像：一张小男孩与其宠物狗的温馨合影。由此，赛尚也被业界誉为“数码相机之父”。

伊士曼柯达公司将这台相机于 1977 年向美国提交专利申请，并于 1978 年获得名称为“电子静态照相机”的美国 US4131919 号专利权，如图 21-1、图 21-2 所示，其发明人为赛尚（Steven J. Sasson）和加雷思（Gareth A. Lloyd），这也是世界上第一台数码相机的专利申请。该发明的核心技术包括电荷耦合器件、电荷转移电路、高速模拟 - 数字转换器、数字缓冲存储器和音频磁带。该电子静态照相机采用音频磁带记录捕获的场景图像，该图像通过从磁带读取所记录的信息，并将其转换为与电视信号接收电路兼容的格式，从而可以在电视机上将其显示。

从专利文献中我们发现，在赛尚的这件美国 US4131919 号专利

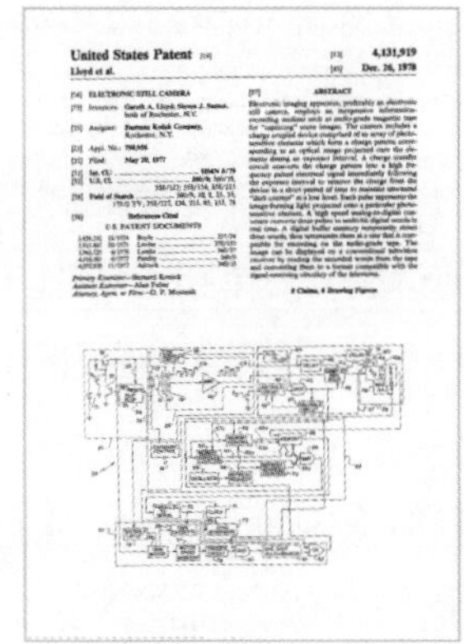

United States Patent 4,131,919
Lloyd et al. Dec. 26, 1978

图 21-1

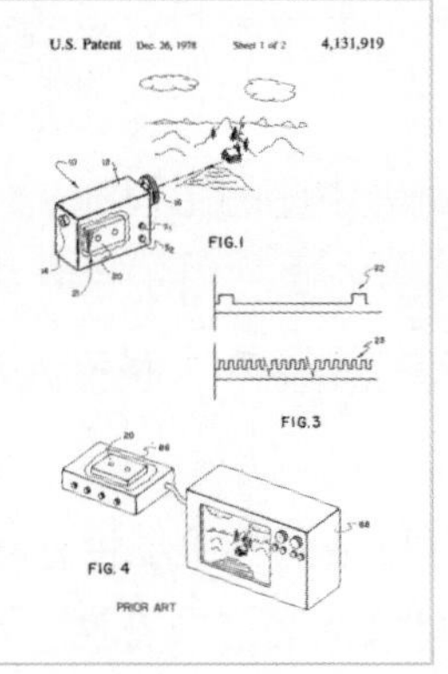

图 21-2

申请提交时已有 5 件美国在先申请，其中之一就是该公司 1974 年提交的名为“用预先记录的定时信号作为控制视频时间基准的磁带便携式摄像机”的专利，于 1976 年获得美国 US3962725 号专利权，如图 21-3 所示。该专利发明人为詹姆斯（James U. Lemke）和罗伯特（Robert A. Letz），其发明核心技术是：摄像机使用磁带记录图像，磁带驱动器很少或不进行调速，摄像机的磁带通过预先记录的信号进行检测扫描从而进行速率控制，摄像机的播放同样由预先记录的信号进行控制从而进行磁带播放。

同时，我们还发现，在赛尚发明之后有 120 多件专利申请参考了赛尚的美国 US4131919 号专利技术，其中包括该公司 1979 年提交的名为“慢帧视频摄像机 / 录音机和图像传感信号处理装置”专利，于

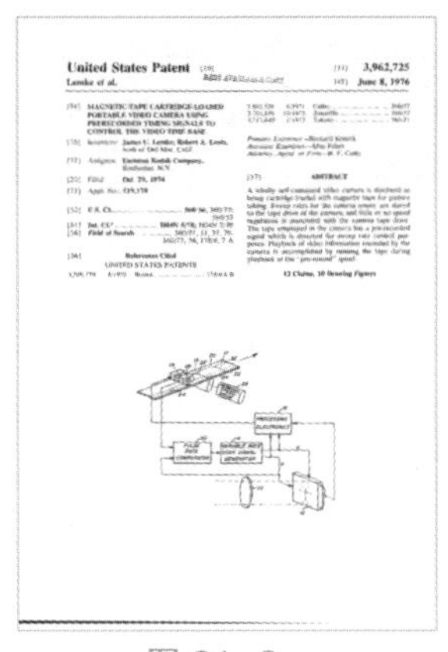

United States Patent 3,962,725
Lemke et al. June 8, 1976

图 21-3

1981 年获得美国 US4263623 号专利权，如图 21-4、图 21-5 所示。该专利发明人为恩颐（Nea-Yea Woo）和埃文 A. 爱德华（Evan A. Edwards），其发明的核心技术是：生成标准电视信号速率的视频信号，使用慢帧信号进行录制和播放，从而降低这台摄像机 / 录音机的带宽要求。

专利文献完整的记录了从 20 世纪 70 年代初至 80 年代末，柯达实验室所产生的 1000 多项与数码相机有关的专利，为数码相机的架构和发展奠定了基础，让数码相机一步步走向现实。1989 年，柯达终于推出了第一台商品化的数码相机。

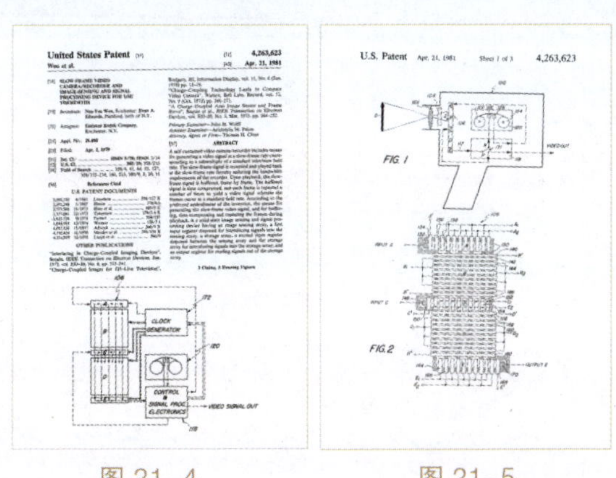
United States Patent 4,263,623
Woo et al. Apr. 21, 1981

U.S. Patent Apr. 21, 1981 Sheet 1 of 3 4,263,623
FIG. 1
FIG. 2

图 21-4　　图 21-5

当第一台数码相机原型机第一次展示给投资者时，人们询问这种产品何时可以成为消费品？赛尚的回答是：大概需要 15 ~ 20 年这种产品才会走进普通消费者家庭。赛尚的判断相当准确，数码相机的发展是一条漫长的道路。可以看到从 1974 年赛尚开始研发第一台数码相机原型机到 1989 年数码相机第一次商品化共用了 15 年的时间，而之后的功能升级周期则越来越短。

随着软件技术的引入，数码相机的功能升级变得易如反掌，达到了日新月异的程度。而芯片技术的进步，也使数码相机的功能越来越多，处理速度越来越快，体积做得越来越小。数码相机极大的增进了拍摄体验，提高了成像的便利性，因此，已成为世界上最受追捧的消费类电子产品之一。

美国 America

1978.US 4131919

——电子静态照相机

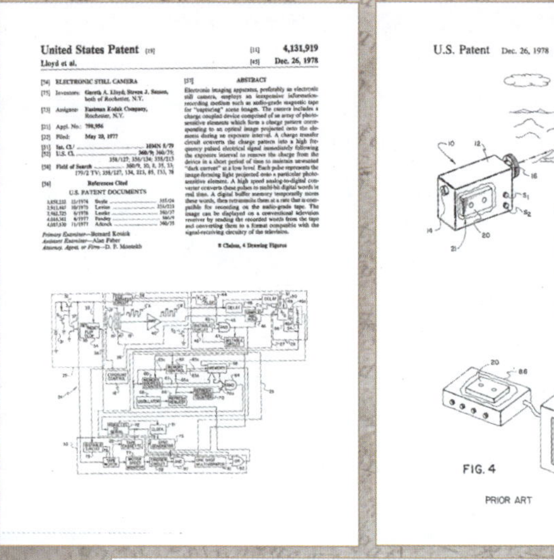

United States Patent [19] [11] 4,131,919

Lloyd et al. [45] Dec. 26, 1978

[54] ELECTRONIC STILL CAMERA

[75] Inventors: Gareth A. Lloyd; Steven J. Sasson, both of Rochester, N.Y.

[73] Assignee: Eastman Kodak Company, Rochester, N.Y.

[22] Filed: May 20, 1977

[56] References Cited

U.S. PATENT DOCUMENTS

Primary Examiner—Bernard Konick

Assistant Examiner—Alan Faber

Attorney, Agent, or Firm—D. P. Monteith

[57] ABSTRACT

Electronic imaging apparatus, preferably an electronic still camera, employs an inexpensive information-recording medium such as audio-grade magnetic tape for "capturing" scene images. The camera includes a charge coupled device comprised of an array of photosensitive elements which form a charge pattern corresponding to an optical image projected onto the elements during an exposure interval. A charge transfer circuit converts the charge pattern into a high frequency pulsed electrical signal immediately following the exposure interval to remove the charge from the device in a short period of time to maintain unwanted "dark current" at a low level. Each pulse represents the image-forming light projected onto a particular photosensitive element. A high speed analog-to-digital converter converts these pulses to multi-bit digital words in real time. A digital buffer memory temporarily stores these words, then retransmits them at a rate that is compatible for recording on the audio-grade tape. The image can be displayed on a conventional television receiver by reading the recorded words from the tape and converting them to a format compatible with the signal-receiving circuitry of the television.

8 Claims, 4 Drawing Figures

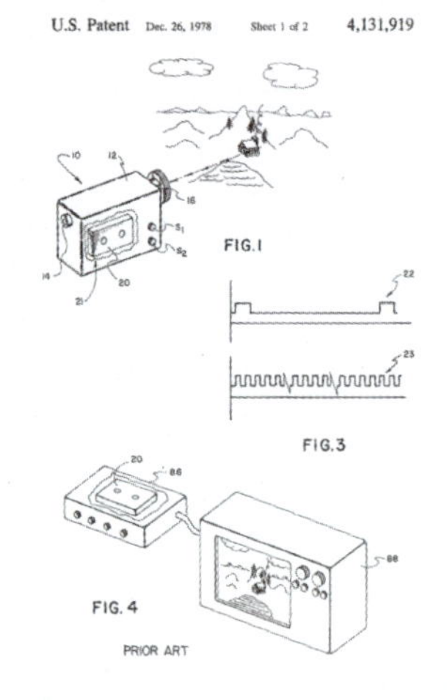

U.S. Patent Dec. 26, 1978 Sheet 1 of 2 4,131,919

U.S. Patent Dec. 26, 1978 Sheet 2 of 2 4,131,919

FIG. 2

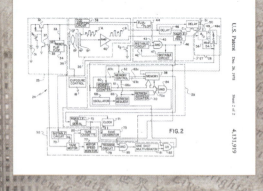

美国

America

ELECTRONIC STILL CAMERA

BACKGROUND OF THE INVENTION

Field of the Invention

This invention relates in general to electronic imaging apparatus and in particular, to an electronic still camera that employs a non-volatile reusable storage medium for recording scene images.

Description of the Prior Art

Conventional cameras employ a shutter for exposing for a limited duration photographic film located at a film exposure plane. Film processing solutions are required to produce a visible image of trhe photographed scene. As is well known, processed photographic film may not be reused.

Recently, considerable effort has been given to the development of solid-state elements for imaging purposes. Such elements offer an advantage over photographic film in that, theoretically, they can be reused any number of times for imaging.

A solid-state imaging element of the type known as a charge transfer device can include a semiconductive substrate doped with majority carriers and covered with a thin insulating film upon which is located an array of electrically conductive electrodes. A depletion region or potential well is formed within the semiconductor under electrodes biased by a voltage of the proper polarity. The use of a charge transfer device involves the basic concept of forming a charge pattern consisting of packets of minority carriers in the potential wells. In the course of an optical imaging process, photons incident upon the semiconductor generate minority carriers within these potential wells in proportion to the amount of light impinging upon the semiconductor in the near vicinity of each well. These packets can be transferred through the semiconductor by sequentially biasing the electrodes. The potential wells effectively "move" through the semiconductor sweeping the minority carriers along within the wells. At an appropriate location these packets may be detected, for example, by removing them sequentially from the semiconductor by means of a reverse-biased diode coupled to transfer electrodes. An arrangement for read-out of information from a charge transfer device is disclosed in *IEEE Transactions On Electron Devices*, Vol. ED-20, No. 6, June 1973, in an article entitled "Interlacing In Charge Coupled Imaging Devices", by Carlo H. Sequin.

As with conventional photographic film, to obtain a scene image having a proper contrast, the imaging element must not be overexposed, or, in other words, the potential wells must not be saturated with minority carriers during the exposure interval. However, even in the absence of illumination, the regions constituting the potential wells tend to saturate with the passage of time by means of the thermal generation of minority carriers. Carriers generated in this manner constitute an unwanted signal commonly known as a "dark current". It is important that this signal be only a small fraction of the signal produced by incident illumination, particularly if the dark current is non-uniform over the imaging area.

U.K. Pat. No. 1,440,792, entitled ELECTRONIC STILL PICTURE CAMERA, and U.S. Pat. No. 4,057,830 which corresponds thereto and is entitled Electronic Photography System, disclose a camera for electronically recording "stop-action" or still pictures

that includes a charge transfer device, and recording apparatus that employs an inexpensive information-recording medium which is non-volatile and reusable, such as a magnetic tape, disc or drum. The camera also includes a conventional shutter mechanism for exposing the transfer device to reflected scene light for a duration related to scene brightness. The rate of read-out of the signals produced by the charge transfer device is synchronized with the speed of the recording apparatus since the transfer device output is connected directly to the input of the recording apparatus. These signals are read-out at a relatively slow speed to record a scene image on the storage medium that is employed. It takes approximately one second to output scene information from the charge transfer device. For "stop-action" photography an exposure interval of approximately 1/20 second or less is needed. Accordingly, the charge transfer device is used both for imaging and until scene information is read-out, for data storage.

That camera suffers from the disadvantage that the charge pattern related to the incident illumination will be adversely affected by thermally generated minority carriers. Not only would it be expected that significant dark current would be produced with a 1-second storage interval, but it could also be expected that the "dark current" would be nonuniform. This is because the storage interval for any potential well, and accordingly the number of thermally generated minority carriers in that well, is dependent upon whether or not that potential well is among the first or the last to sweep through the semiconductor to an output transfer electrode. Furthermore, saturation of some potential wells may occur if too many minority carriers are thermally generated. Excess minority carriers would spread to adjacent potential wells to be added to minority carriers in nonsaturated potential wells.

SUMMARY OF THE INVENTION

An object of the present invention is to provide improved electronic imaging apparatus.

In accordance with the above object, the present invention is addressed to an electronic still camera which includes a solid-state imaging device that produces a charge pattern corresponding to an optical image projected onto an imaging surface during an exposure interval. Charge transfer means converts such charge pattern into a high frequency pulsed electrical signal within a relatively brief time after termination of the exposure interval. Electrical signal transforming means receives data corresponding to this electrical signal in real time, then retransmits such data at a substantially slower rate to recording apparatus. This slower data rate permits recording of signals corresponding to the optical image on an inexpensive recording medium such as audiograde magnetic tape.

The invention, and its objects and advantages, will become more apparent in the detailed description of a preferred embodiment presented below.

BRIEF DESCRIPTION OF THE DRAWINGS

In the detailed description of a preferred embodiment of the invention presented below, reference is made to the accompanying drawings in which:

FIG. 1 pictorially represents an electronic still camera in accordance with the teachings of the present invention;

FIG. 2 is a schematic block diagram of a solid-state image sensor, and signal processing circuitry in the camera of FIG. 1;

FIG. 3 is a drawing of waveforms to help explain the operation of the solid-state image sensor and signal processing circuitry of FIG. 2; and

FIG. 4 is a perspective view of apparatus for displaying pictures of scenes recorded by the camera of FIG. 1.

DESCRIPTION OF A PREFERRED EMBODIMENT

The present description will be directed in particular to elements forming part of, or cooperating more directly with, apparatus in accordance with the present invention. It is to be understood elements not specifically shown or described may take various forms well known to those having skill in the art.

There is shown in FIG. 1 of the drawings an electronic still camera, designated generally by the numeral 10. As with conventional still cameras of the type that employ photographic film for "capturing" light from a scene being photographed, camera 10 includes a housing 12, a viewfinder 14, and a taking lens 15 (shown in FIG. 2) located within a lens barrel 16. A diaphragm mechanism 17 (shown in FIG. 2), coupled to an exposure control circuit 18 (also shown in FIG. 2), forms a variable aperture aligned with lens 15 to control the amount of scene light transmitted through the lens. Diaphragm control is automatic and is provided by means, well known in the art, that controls the lens aperture in accordance with the intensity of reflected scene light, the control being effected by the current produced by a photoconductive element, such as a photocell, of circuit 18. Photoconductively controlled diaphragm mechanisms are well known in the art and are disclosed in several patents, one of which is U.S. Pat. No. 3,962,707 filed in the name of F. M. Galbraith, Jr. and commonly assigned herewith.

Unlike conventional still cameras that employ photographic film, camera 10 employs a solid-state image sensor 19 (shown in FIG. 2), located at the focal plane of lens 15, and a record storage medium, such as, for example, reusable magnetic tape, for recording electrical signals corresponding to scene images projected onto the imaging surface of sensor 19. The tape may, for example, be located in a magnetic tape cassette 20, which is located in a cassette-receiving chamber 21. Chamber 21 is accessible through a camera door (not shown) and includes means (also not shown) for removably mounting the cassette 20 in the chamber. Cassette 20 is adapted to be connected to suitable data retrieval apparatus, as referred to hereinafter, to obtain visual displays of recorded scenes.

Sensor 19 constitutes a charge transfer device and, in particular, is a charge coupled device (CCD), which produces a pattern of charge carriers that is an analog representation of an optical image focused onto an imaging surface of the CCD. As is known in the art, a CCD may include a silicon substrate covered by oxidized silicon, with an array of closely spaced conducting pads on the silicon dioxide surface. The pads serve as gate electrodes and may, for example, be formed by selectively doping a layer of transparent polysilicon material. These gate electrodes are interconnected in rows (or columns), which are electrically connected to both a CCD address circuit for sequentially applying gate electrode voltages, and to a data readout register.

When a CCD is used as an imaging device, charge carriers are produced by light quanta absorbed in the silicon, the number of carriers being proportional to the amount of radiant energy reaching the silicon. During the so-called "integration time", comparable to the exposure time in conventional camera art, the CCD address circuit applies voltages to the gate electrode array to provide a pattern of photosites having potential wells which attract charged minority carriers. A charge pattern will form in the potential wells under the various photosites nearest to where the charges are produced. The number of charge carriers which accumulate in each potential well during the integration time is proportional to the amount of radiant energy reaching that well and this, in turn, is proportional to the radiation intensity and the duration of the integration time. Thus, a spatial pattern of carriers corresponding to an optical image is produced by the CCD. For further, more detailed descriptions of CCD's, reference is made to one of the numerous publications on such devices, U.S. Pat. No. 3,858,232, entitled INFORMATION STORAGE DEVICES, the disclosure of which is incorporated herein by reference.

By the proper manipulation of voltage potentials to the interconnected rows (or columns) of electrodes following this integration time, this charge pattern can be shifted to one point of the CCD, where, by means of a suitable output connection the charge collected can be removed sequentially into the aforementioned data readout register. This stream of charge then constitutes an electrical current which is an analog replica of the light projected onto the CCD. Such a current is caused to flow through an output load to generate a train of voltage pulses, each pulse representing one small area of the projected image with pulse amplitude being proportional to the amount of radiation reaching a particular photosite.

With our presently preferred CCD apparatus, an image is formed in two separate, interlaced fields, each field containing one-half the total number of photosites. Photosites are formed in the first field for 25 milliseconds (ms), then photosites are formed in the second field for the next 25ms interval, the total "exposure interval" being 50ms. While charge packets are formed in the second field, the charge pattern in the first field is read out and converted to a train of voltage pulses, the pulse repetition rate being inversely proportional to the interval used for read-out. For detailed descriptions of CCD interlacing techniques and apparatus, reference is made to U.S. Pat. No. 3,911,467, entitled INTERLACED READOUT OF CHARGE STORED IN CHARGE-COUPLED IMAGE SENSING ARRAY, and to the aforementioned article by C. H. Sequin.

Since the fields are interlaced, scene information is reorganized so as not to be presented for display in the same interlaced timing format. Apparatus for data reorganization is provided by the aforementioned data retrieval apparatus.

A type 201 CCD, manufactured by the Fairchild Camera and Instrument Company (FCI), includes a 100 × 100 photosite array. For that CCD, an optical image may be represented by a CCD output pulse train of 10-thousand pulses. Four pulses of such a train are denoted A in FIG. 2.

Two of the clock signals associated with CCD 19 are illustrated schematically in FIG. 3. Waveform 22 constitutes a master clock signal and is used to signal the beginning and end of a pulse train that represents an

image projected onto the CCD during one "exposure interval", e.g. 50ms. For a pulse train duration of, for example, 50ms, waveform 22 would be a 20 cycle per second (Hz) signal. (The duration of such a pulse train and an exposure interval need not be the same.) Waveform 23 is a high-frequency signal that is used to synchronize the operation of signal processing circuitry, denoted generally 24, with the occurrence of each pulse in the CCD output signal, as explained in detail hereinafter. If 10,000 pulses are produced in a 50ms interval, waveform 23 would be a 200-thousand Hz signal.

In accordance with our invention, light from a scene to be recorded is "captured" electronically in the following manner. A camera operator first actuates a power control member to apply electrical power to circuitry 24. At that instant, CCD address circuits cause the CCD to commence producing its pulsed output signals, a pulse train being produced for each exposure interval. In addition, light-sensitive exposure control circuit 18 positions diaphragm 17 to form a lens aperture in accordance with the level of ambient light. No CCD output pulse train is processed, however, until a second control member is actuated. Once such actuation occurs, a CCD output signal, representing an optical image projected onto an imaging surface during one exposure interval, is processed. Each CCD pulse for the interval is digitized in real time to form a digital word, i.e., a total of 10,000 digital words for the aforementioned type 201 CCD, and such words are advantageously stored in a high-speed buffer memory. These digital words are sequentially retransmitted, preferably, after the buffer memory receives the last digital word, at a rate that is substantially lower than the CCD pulse output rate. This low signal rate permits inexpensive recording apparatus to be used for recording scene information.

With our presently preferred camera apparatus, the scene information for the first field is stored in the buffer memory during the time the charge pattern for the second field is formed. Accordingly, the total time used to "capture" a single scene image and store it in the buffer memory is approximately 75ms. Scene information is transferred from the buffer to the recording apparatus in approximately 23 seconds.

With this arrangement, CCD dark current is maintained at a low level since the CCD is not employed for information storage in the same sense as is the imaging device employed in the electronic camera disclosed in the aforementioned U.K. patent 1,440,792. Furthermore, there is no need to employ expensive recording apparatus, such as broadband video apparatus. In fact, recording apparatus having relatively slow recording or data-write speeds, such as for example, audio-grade magnetic recording apparatus, may be utilized to reliably and accurately record information transmitted from the buffer memory.

It shall be understood that an "integration interval" and the time used to output scene data from the CCD for one field need not be predetermined nor of the same duration. For example, an integration interval and/or exposure interval could be controlled by a light-sensitive integrating circuit, the control being effected by the time required to charge a capacitor to a predetermined level by a current through a photoresistive element. The capacitor and photoresistor constitute an integrating circuit having a variable time constant related to the intensity of the light impinging on the photoresistor. Read-out of scene data from the CCD could, for exam-

ple, commence in timed relation to the capacitor being charged to such level. Photoconductively controlled timing circuits are well known in the art and are disclosed in many patents, one of which is U.S. Pat. No. 3,672,267, entitled SEQUENTIAL CONTROL FOR CAMERA DIAPHRAGM AND SHUTTER, and commonly assigned herewith.

An important feature of our invention is, however, that read-out of the charge pattern for each field occurs rapidly so that the CCD is not employed for any significant period of time as a storage device.

Referring now to FIG. 2, there is shown a schematic diagram of signal processing circuitry 24 for obtaining still pictures of scenes imaged onto CCD 19. Circuit 24 includes generally a power control circuit 25, and image-sensor apparatus 26, including CCD 19, for producing a high frequency pulsed electrical signal corresponding to an imaged scene, each pulse in the signal having an amplitude proportional to the amount of light impinging on a particular photosite of the CCD. A circuit 27 constitutes an electronic shutter to control the signal produced by CCD 19 that is to be recorded. An analog-to-digital (A/D) converter circuit 28 converts in real time the signal transmitted by circuit 27 into multi-bit digital words. Buffer circuitry 29 functions to receive and temporarily store such words in real time, then transmits them at a rate that is substantially lower than the real time rate at which these words are loaded into the buffer. Circuit 24 also includes recording apparatus 30 having a data recording speed that is compatible with the rate that data is transmitted from buffer 29.

As with a conventional still camera that employs photographic film to obtain a "photograph" lens barrel 16 (FIG. 1) is pointed at a scene visible through viewfinder 14. Electrical power is provided for circuitry 24 from a battery 31 upon actuation of a momentary, pushbutton switch S_1 mounted on housing 12, as shown in FIG. 1. When switch S_1 is closed, the baseemitter junction of transistor 32 is forwardly biased to turn the transistor ON, and collector current flows to turn signal processing circuitry 24 ON. At the same time, a bias signal is produced on conductor 33 that is applied to a set-reset flip flop 34. When this happens, flip flop 34 assumes a set condition and produces at its output 35 a high-level voltage to thereby turn transistor 37 ON. When a voltage signal above a predetermined level is applied to its input 36, flip flop 34 switches into its reset condition and its output 35 produces a low-level voltage to turn transistor 37 OFF. When this occurs, with switch S_1 in its open position, transistor 32 is turned OFF to remove electrical power from circuitry 24.

Image-sensor apparatus 26 includes a CCD address circuit 38, which causes signals 22 and 23 in the form of clock pulses to be applied to CCD gate electrodes in a known manner to produce the aforementioned pulsed electrical output signal. This output signal is produced during a brief interval immediately following the integration time for each of the aforementioned two fields so that dark current is maintained at a low level. In a preferred embodiment, FCI type 201 CCD is employed using a 25ms integration time and 25ms readout interval for each field, the total readout interval being 50ms for the signal representing an optical image.

Image-sensor apparatus 26 also includes a video amplifier 40 to filter out noise in the CCD output signal, as illustrated diagrammatically by the signal waveform B at the output of amplifier 40. Since CCD's typically are sensitive to infrared (IR) radiation, image-sensor appa-

ratus 26 may further include an IR filter 41 interposed along the optical axis of camera 10 between lens 15 and CCD 19. Filter 41 blocks IR radiation so the charge pattern produced by CCD 19 is more closely related to impinging radiation in the visible spectrum. Factors affecting CCD performance are discussed in *IEEE Transactions On Electron Devices*, Vol. ED-18, No. 11, November 1971, page 986, entitled "Charge-Coupled Imaging Devices: Design Considerations", by G. F. Amelio et al.

As soon as electrical power is applied to circuit 24, CCD 19 immediately begins to produce its pulsed electrical output signals. For successive 50ms exposure intervals families of pulse trains are generated with each pulse train representing the optical image formed during one such interval. Until momentary, pushbutton switch S_2, mounted on camera 10, is depressed, however, circuit 27 is effectively disabled, and image data is not transmitted to apparatus 30 for recording. Circuit 27, in effect, operates as an electronic shutter or gate. However, unlike a conventional camera shutter which operates to control the time during which light is allowed to reach a light-sensitive medium, i.e. film, this electronic shutter controls the signal, representing an optical image, that is to be recorded. The reason circuit 27 is effectively disabled immediately after power is applied to circuitry 24 is to permit CCD 19 to clear itself of thermal charge prior to recording an image signal. With out presently preferred CCD apparatus, we have found that approximately one-half second is needed to clear the CCD of such charge after power is applied.

Circuit 27 operates as follows: Momentary closure of switch S_2 causes a bistable circuit 42 to produce a logic "1" signal at its output 43. A flip flop 44 produces both logic "1" and "0" signals at its output following the occurrence of odd numbered and even numbered pulses, respectively, in the aforementioned 20Hz signal 22. Accordingly, an AND gate 45, which conducts in response to logic "1" signals at each of its inputs, produces a logic "1" output signal following the occurrence of the first odd numbered pulse in signal 22, after switch S_2 is actuated. The logic "1" signal turns ON a conventional sample-and-hold circuit 46. Circuit 46 serves to extend the duration of each pulse in waveform B for the period between the trailing edge of one pulse and the leading edge of the next pulse. Accordingly, sample-and-hold circuit 46 produces an electrical signal having a series of contiguous pulses. Such a signal is denoted by the letter C in FIG. 2. When the output signal produced by gate 45 returns to a logic "0" state, such as when the next pulse (an even numbered pulse) in signal 22 is applied to flip flop 44, circuit 46 is turned OFF. At that time, however, 50ms have passed and an electrical signal representative of a full scene image has been transmitted through sample-and-hold circuit 46.

When the output of gate 45 is a logic "0", a logic "1" signal occurs at the output of a bistable circuit 47. This logic "1" operates to reset circuit 42 so that its output returns to its normal, logic "0" state. This is done to prevent the output of gate 45 re-enabling circuit 46 upon the occurrence of the next odd numbered pulse in waveform 22, without the reactuation of switch S_2.

The aforementioned high frequency signal 23 from CCD address circuit 38 is used to synchronize the operation of sample-and-hold circuit 46 with waveform B. A delay circuit 48 delays signal 23 to allow for CCD readout and signal delay through amplifier 40. Signal 23 is

delayed by circuit 48 to synchronize the operation of sample-and-hold circuit 46 with the occurrence of each pulse in waveform B so that pulses are extended at the proper point in time and with the correct amplitude.

Waveform C is applied to A/D converter 28 which converts each extended pulse into a 4-bit digital word having a magnitude representative of pulse amplitude. Converter 49 produces these bits sequentially, the most significant bit is produced first on output 49a, and the least significant bit is produced last on output 49d. A start-conversion signal is supplied on conductor 50 to permit converter 49 to commence each signal conversion operation in synchronism with the occurrence of each waveform C pulse. This is done by means of signal 23 which is delayed a predetermined amount by a delay circuit 51.

A digital latch 54 is provided to ensure that the bits representing each digital word are not changing during signal transmittal to buffer circuitry 29. A/D converter 28 produces an end-of-conversion (EOC) signal on conductor 56 once each least significant bit is produced, to signal latch 54 to transmit a digital word to buffer 29.

Buffer circuitry 29 constitutes means for receiving signals in the form of digital words from A/D converter 28 in real time, and transmitting such words at a rate that is substantially lower than the rate at which these words are received from the A/D converter. Since digital words representing a single optical image are loaded into memory 58 within 50ms, while approximately 23 seconds are used to transmit such words from the memory to recording apparatus 30, buffer 29 operates to reduce by a factor of approximately 460 the rate that digital words are transmitted to apparatus 30 from the buffer compared to the rate such words are transmitted from the CCD into the buffer. To accomplish this, the buffer 29 includes a highspeed digital memory 58 for temporarily storing words as they are accumulated prior to transmittal from the buffer to recording apparatus 30. Preferably, buffer 29 is operated to transmit stored words after memory 58 has received the last word. This is done to simplify the operation of buffer 29 and does not operate to significantly increase the time between when the signal is produced by CCD 19 and when such signal is recorded by apparatus 30 since all words are loaded into memory 58 within 75ms after an exposure interval commences. It shall be understood, however, that it is within the spirit and scope of our invention that signal transferral from memory 58 can commence as soon as the first digital word is loaded into memory.

Buffer 29 operates as follows: The aforementioned EOC signal produced on conductor 56 is also applied to a memory address counter 60 included within buffer 29. Counter 60 serves a record keeping function for memory 58 by providing an addressenable signal along input address line 61 to control where in memory either write or read operations are to occur. Address counter 60 is initialized to ZERO when AND gate 45 switches its output either from a logic "0" to a logic "1" or from a logic "1" to a logic "0". When an EOC signal is produced, counter 60 increments its count by one to permit the memory location corresponding to that count or address to receive a digital word for storage.

A memory control circuit 62 provides command signals to memory 58 to control whether information is to be written into a read from memory. Control circuit 62 is enabled to produce memory write-control and read-control signals on conductor 63a when the output

of gate 45 is a logic "1" and a logic "0", respectively. These write-control and read-control signals turn input write-logic and read-logic gates (not shown) ON in memory 58 to permit data to be written or read, respectively, from memory at the address corresponding to the count provided by counter 60. Such write-control and read-control signals are produced when data-ready signals are provided on input conductor 63c. These data-ready signals occur either when the aforementioned EOC signals are produced, or when recording apparatus 30 produces its output signals, as explained in detail hereinafter.

Memory 58 is comprised of a plurality of solid-state random access memories (RAM's) which are connected together to provide high-speed storage for 10,000 4-bit words within the aforementioned 50ms interval. Solid-state memories have the speed for storing data this quickly. However, as is known in the art, solid-state RAM's may periodically require "rejuvenation" or "refreshing" to maintain data as originally stored. Buffer 29 therefore, includes memory refresh circuitry comprising a refresh-request circuit 66, under the control of an oscillator 68, and a refresh address counter 70. Circuit 66 produces a refresh-request signal on conductor 66a, and counter 70 produces a refresh-address signal on conductor 70a. Each of such signals are produced at 63 microsecond intervals. The refresh-address signal determines which memory locations are to be refreshed. Each refresh-request signal causes memory control 62 to produce a memory refresh signal on conductor 63b to refresh the data at the memory address corresponding to the count provided by counter 70.

The circuitry within memory 58 by which stored information is refreshed, and information is stored and/or written into the memory is readily available commercially. In our preferred embodiment, memory 58 is comprised of dynamic RAM's manufactured by the Motorola Corporation, and designated MCM6605.

An information storage device that requires electrical power to retain or preserve stored information presents obvious advantages, particularly for camera users who wish to store recorded information for long intervals. Accordingly, it is desirable to provide a storage device by which scene information may be permanently stored or recorded without the need for electrical power to preserve the information once it is recorded. A storage device of this type is known in the art as non-volatile. In addition, it is desired that such a storage device be inexpensive. To meet these objectives, buffer memory 29 includes circuitry for outputting stored data at a rate below the 200 KHz rate, e.g. 10,000 words within the 50ms interval, that data is received. The desired data output rate is controlled so that it is compatible with the recording speed of a permanent inexpensive recording apparatus, such as, for example, an audiograde magnetic tape recorder. Such circuitry controls the rate data is read from memory 58 and includes a clock 71, a parallel-to-serial bit converter 72, and sync-pulse generator 73. Generator 73 is controlled by clock 71 and produces at its output 75 both a sync signal to be recorded and a data-ready signal to read stored information from the memory.

Recording apparatus 30 includes, in addition to cassette 20, a tape recorder motor 76, a motor speed monitor circuit 77, a voltage-sensitive trigger circuit 78, and a bistable circuit 79. When switch S_2 is closed, bistable circuit 79 produces a logic "1" signal at its output to turn motor 76 ON to drive the tape in cassette 20. Data

commences to be read from memory 58 to be recorded on the tape once both tape recorder motor 76 is operating above a predetermined speed, and the last digital word has been stored in buffer 29. Since motor 76 does not reach such speed instantaneously following the actuation of switch S_2, motor speed monitoring circuit 77 is provided to produce a voltage signal having an amplitude proportional to motor speed. This voltage signal is applied to the input of voltage-sensitive trigger circuit 78. Circuit 78 is of the type known in the art as a Schmitt trigger circuit that produces a low-level voltage at its output when a voltage is present at its input that is less than a predetermined level. Likewise, when a voltage signal applied to its input exceeds a predetermined threshold, a high-level voltage signal is produced at its output that is applied to one of two inputs to an AND gate 80.

In addition, when AND gate 45 switches its output from a logic "1" to a logic "0", indicative of when a full scene image has been transmitted through circuit 46 and, accordingly, into buffer 29, a high-level signal is applied to the second of the two inputs to AND gate 80. It would be necessary to invert the logic "0" signal at the output of gate 45 if such signal were a low-level signal. This could, of course, be done by means of an inverter circuit (not shown).

AND gate 80 conducts in response to high-level signals at each of its inputs. Accordingly, gate 80 produces an output signal once motor 76 is operating above a predetermined speed, and a full scene image has been loaded into buffer 29.

When AND gate 45 switches its output from a logic "1" to a logic "0", control circuit 62 is enabled to produce the aforementioned read-control signal when a data-ready signal occurs on conductor 63c. In a manner similar to when a write-control signal is produced, circuit 62 produces voltage signals to set input read-logic gates (not shown) to permit data to be read from memory 58.

Recording apparatus 30 begins to record data stored in buffer 29 once AND gate 80 produces its ouput signal. Such signal causes a one-shot multivibrator 81 to produce an output pulse that is applied to OR gate 82. When this occurs, memory address counter 60 is caused to increment its count by 1. In addition, a data-ready signal is applied to control circuit 62 along conductor 63c. This signal causes control circuit 62 to produce a read-control signal which is applied to memory 58 along conductor 63c. When this happens, a 4-bit word is read out of memory 58 in parallel bit form, at the address determined by counter 60, and is applied to parallel-to-serial bit converter 72. Each 4-bit word is converted to a 4-bit word in serial form and transmitted to tape cassette 20 for recording. At the same time, sync-pulse generator 73 produces a sync bit which is added to the 4 serial bits to produce a 5-bit word that is recorded on magnetic tape loaded in cassette 20.

Each sync pulse is also applied to OR gate 82 to initiate a subsequent read-control signal and to increment address counter 60 by one. The preceding data read sequence continues until each of the 10,000 4-bit words are read from memory and recorded on magnetic tape. Once the last word is recorded, OR gate 82 increments counter 60 to 10,001. When this occurs, and AND gate 84 produces a high-level voltage signal that is applied to input 36 of flip flop 34. Accordingly, flip flop 34 produces a low-level voltage at its output 35 and transistor 37 turns OFF, thereby removing electrical

power from camera 10, by automatically turning camera 10 OFF, inadvertent power drain from camera battery 31 is prevented.

Additional pictures are recorded by sequentially reactuating buttons S_1, S_2. However, cassette 20 may be removed from camera 10 after any number of scenes have been recorded.

In a presently preferred embodiment tape recorder apparatus 30 records scene information at a data rate of approximately 2,150 bits/second. Since 5 bits are used to represent the scene light sensed at each of 10,000 photosites (4 information bits and 1 sync bit) 50,000 total bits are used to record each picture. Consequently, an electrical signal representative of a single scene is read from memory 58 and is magnetically recorded in approximately the aforementioned 23 seconds. The recording rate of 2,150 bits/second resulted from a recording density of 390 bits/inch on the tape and a maximum tape speed of 5.5 inches/second. There are, of course, other audio-grade recording systems that would utilize tape having an inherently higher recording density and/or a higher tape speed and which would accordingly reduce the time for magnetically recording a scene. In addition, source and/or channel encoding schemes could be employed to more efficiently record information on the tape. With our presently preferred audiograde tape recorder, up to 25 pictures can be recorded on a single cassette tape.

There is shown in FIG. 4 an arrangement by which recorded scene information may be displayed for viewing. Such an arrangement includes a microcomputer 86 of the type manufactured by the Motorola Corporation, and a conventional TV monitor 88. Microcomputer 86 reads scene data from cassette 20 and temporarily stores it in an internal memory (not shown). Microcomputer 86 also reorganizes the data so that it is not presented for display with the same interlaced timing format that data is read out of CCD 19. Digital words are converted to analog form and arranged in a TV-scan format for display on monitor 88. Apparatus, such as microcomputer 86, for converting information data to a format suitable for TV display are well known in the art.

One advantage of the arrangement shown in FIG. 4 is that the camera operator may first visually display each recorded scene, then decide whether to keep the picture. Unwanted scenes may be erased from the tape, and the tape may be reused without incurring additional cost.

Permanent photographic prints may be made on conventional photographic paper. This can be done using electrooptical signal translating apparatus. For example, scene information may be read from magnetic tape and reconverted into a pulsed electrical signal, such as the aforementioned waveform C. Individual pulses in waveform C would then be applied to timing circuitry for controlling the time duration that photographic paper would be exposed. Since each pulse is related to the brightness of a particular point in the recorded scene, the photographic paper may be exposed by a series of variable-duration light signals from a light beam scanned across the paper.

While we have described our invention as employing a magnetic tape for low cost non-volatile information storage, it shall be understood that other non-volatile informationstorage media may be utilized. We prefer to use a non-volatile storage device for user convenience and to reduce power consumption. One other storage medium that appears very attractive for meeting these

criteria is a magnetic bubble device. This type of storage device is attractive not only because it is basically non-volatile and requires low power, but offers the further advantage of large storage capacity and may potentially be used to store millions of bits in one device. If this were the case, a magnetic bubble device could replace the magnetic tape recorder.

Although various specific elements and data handling capabilities and characteristics have been used to describe our electronic camera invention, it shall be understood by those skilled in the art that numerous modifications may be made that are within the spirit and scope of our inventive contribution. For example, picture quality could be improved with the use of more bits per photosite to produce improved scene contrast. In an article by A. A. Goldberg, entitled "PCM Encoded MTSC Color Television Subjective Tests", and appearing in *SMPTE*, Vol. 82, No. 8, August 1973, it was reported that no noticeable degradation of picture quality (on an NTSC 525-line, 60-field TV signal) occurred when 6 bits per photosite were used, compared to a picture produced with an undigitized signal.

Furthermore, color pictures are possible using multiple CCD's with appropriate filters or with the development of a single, color responsive CCD. For "capturing" color images, an electrical signal for each of three primary colors could be produced in a manner as set forth hereinbefore. Recorded scene information may be displayed for viewing by transforming each signal into its corresponding color, and forming a composite polychromatic picture. For a detailed description of a solid-state color imaging device, reference is made to U.S. Pat. No. 3,971,065, entitled COLOR IMAGING ARRAY, and assigned to the assignee of the present invention.

Electronic imaging apparatus within the teachings of our invention provides a user the opportunity to take pictures in available light. By utilizing other CCD properties, other advantages can be realized. For example, IR photography is possible with the same camera by simply filtering the light before it reaches the CCD.

In addition, although we have described a preferred embodiment of our invention as constituting an electronic still camera, it shall be understood that the essence of our inventive contribution is also applicable to other electronic imaging apparatus such as, for example, apparatus for recording fast-action scenes such as movie camera apparatus.

We claim:

1. In an electronic still camera having a solidstate light-responsive device for producing at a relatively high rate a stream of discrete signals indicative of optical images received by said device during adjacent exposure intervals, the improvement comprising:
(a) means for extracting said discrete signals from said light-responsive device during a time interval which is no greater than the interval during which said optical images are produced;
(b) circuit means for separating said signals from each other;
(c) means for transforming said separated signals into a stream of signals occurring at a rate which is slower than said high rate; and
(d) means for recording in real time on a non-volatile medium said slower rate stream of signals.

2. Apparatus as set forth in claim 1 wherein said transforming means comprises storage means for enabling production of said high rate stream of signals and re-

cording of said slower rate stream of signals to occur concurrently.

3. In an electronic still camera having a solid-state light-responsive device located to receive an optical image, said light-responsive device including electrical addressing means for producing families of electrical signals relating to charge patterns formed in said light-responsive device during successive exposure intervals, each signal within a family corresponding to a charge pattern formed during one exposure interval and each of said signals defining a first train of information bearing indices occurring at a first relatively high rate, the improvement comprising:
(a) control means for providing a signal to said electrical addressing means to extract said families of electrical signals from said lightresponsive device at a speed which is at least as great as the speed at which said charge patterns are formed;
(b) electrical gating means for effectively isolating at one of said electrical signals from the other of said electrical signals;
(c) means associated with said gating means for transforming the electrical signal which has been isolated into a second train of information bearing indices occurring at a rate which is substantially less than said first rate of said first train of information bearing indices; and
(d) means for recording on a medium said second train of information bearing indices.

4. In an electronic still camera having a solid-state imaging device located at an exposure plane, means associated with said imaging device for forming charge patterns relating to optical images projected onto said imaging device during adjacent exposure intervals, the improvement comprising:
(a) means for producing a stream of discrete electrical signals, representative of said charge patterns, at a speed which is at least as great as the speed at which said charge patterns are formed;
(b) electrically energizable circuit means for separating a one of said electrical signals from the other of said electrical signals;
(c) means responsive to said separated electrical signal for producing a plurality of multi-bit words indicative of said separated electrical signal, said digital words being produced in real time;
(d) data storage means;
(e) means for transferring said digital words into said data storage means in real time at a first relatively high rate;
(f) means for retrieving said digital words from said data storage means at a second rate that is substantially lower than said first relatively high rate; and
(g) recording apparatus for recording said retrieved digital words on a non-volatile recording medium, said recording apparatus having a recording speed that is compatible with said second rate.

5. In an electronic still camera having a light-responsive device located at an exposure plane for producing charge patterns relating to optical images projected onto said light-responsive device during adjacent exposure intervals, the improvement comprising:
(a) control means for providing a signal;
(b) electrical addressing means associated with said light-responsive device and responsive to said signal for extracting from said light-responsive device information-bearing electrical signals corresponding to said charge patterns at a speed which is at

least as great as the speed at which said charge patterns are formed;
(c) electronic shutter control means having (1) a first state for blocking transmittal of said electrical signals, and (2) a second state for transmitting a one of said electrical signals;
(d) actuatable means for switching said control means from its state into its second state;
(e) a signal storage device for receiving and storing in real time at a first data rate representations corresponding to said transmitted electrical signal;
(f) means for retrieving said representations from said signal storage device at a second data rate that is substantially less than said first rate; and
(g) recording apparatus for recording said representations on a non-volatile recording medium, said recording apparatus having a recording speed that is suitable for recording said representations at the rate at which said representations are retrieved from said signal storage device.

6. An electronic still camera as set forth in claim 5 including means for automatically switching said control means from said second state back to said first state following transmittal of said information-bearing electrical signal.

7. In an electronic still camera having a solid-state light-responsive imaging device located to receive optical images, the improvement comprising:
(a) electrical addressing means associated with said imaging device for producing families of electrical signals derived from optical images formed on said imaging device during adjacent exposure intervals, each signal within a family corresponding to an optical image formed during one exposure interval and said signals being produced at a first rate which is at least as great as the rate at which said optical images are formed;
(b) electrical gating means having (1) a first state for blocking transmittal of said electrical signals, and (2) a second state for transmitting said electrical signals;
(c) switching means synchronized with said electrical addressing means for switching said electrical gating means from said first state into said second state then back to said first state for transmitting a given one of said electrical signals;
(d) means for temporarily storing representations of said transmitted electrical signal in real time and for retransmitting said representations at a second rate that is substantially less than said first rate; and
(e) recording apparatus for recording said representations on a non-volatile recording medium, said recording apparatus having a signal access speed that is compatible with the second rate at which said representations are retransmitted.

8. In an electronic still camera including a charge coupled device defining an array of light-sensitive elements located at an exposure plane, the improvement comprising:
(a) electrical addressing means associated with said light-sensitive elements for producing families of pulsed electrical signals relating to optical images projected onto said elements during contiguous exposure intervals, said signals being produced at a rate which is at least as great as the rate at which said optical images are formed;
(b) an electrical gating means having (1) a first state for blocking transmittal of said pulsed electrical

signals, and (2) a second state for transmitting said electrical signals;
(c) actuatable switching means for switching said gating means from said first state into said second state;
(d) said electrical gating means including means for automatically switching said gating means from said second state back to said first state following transmittal of the first one of said electrical signals which occurs after actuation of said switching means;
(e) a digital data storage device;

(f) converter means for converting in real time said pulses in said first one of said electrical signals into a multi-bit digital word;
(g) means for transferring said digital words in real time into said digital data storage device;
(h) means for retrieving said digital words from said data storage device at a rate that is more than an order of magnitude less than the rate at which said words are transferred into said storage device; and
(i) recording apparatus coupled to said digital words retrieving means for recording said digital words on a magnetic recording medium, said recording apparatus having a recording speed that is compatible with recording signals at an audio-grade recording rate.

* * * * *

发条式收音机的发明
Spiral spring type radio invention

英国著名发明家特莱佛·贝利斯（英国发明科学院的创立者）最重要的发明是1994年问世的、无需用交流电或电池的发条式收音机。这一设计还带来其他两大效益：一是当地生产这种收音机的厂家雇用的都是残疾人，二是由于很多穷人都拥有了这种便宜的收音机，政府关于防治艾滋病的广播宣传起到了较为明显的效果。

这一技术的延伸产品是发条式手电筒和发条式电池充电器，都广受欢迎。

GB 2262324

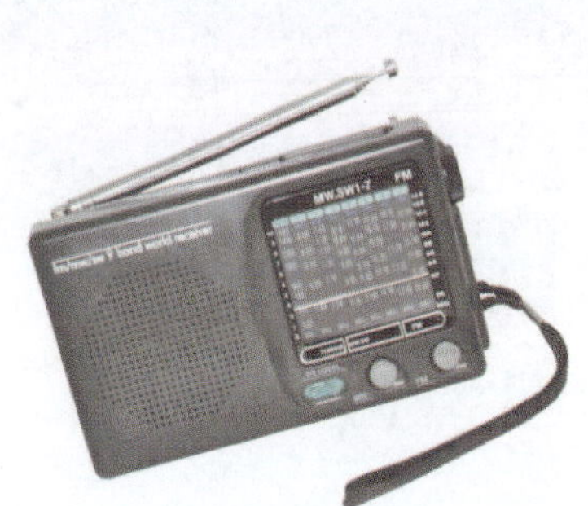

第二十二篇 发条式收音机

随着人类在社会进步的过程中对环境问题的日益重视，绿色低碳已经成为流行的新风尚，节能环保的科技和产品不断涌现，正在影响和改变着我们的生活。手摇发电就是一种绿色的能源供应方式——一种将人体生物能转化成电能的简单方法，今天它的应用包括手电筒、电话、音乐播放机等。而早在 20 世纪 90 年代初使用这种供电方式的消费类电子产品就已经出现在市场上，这就是英国著名发明家特莱佛·贝利斯（Trevor Baylis，1937-）发明的发条式收音机。

特莱佛曾是一名游泳特技表演者，由于职业原因，他对在工作中受伤致残的同行特别关注，为了对残疾人提供帮助，他开始了自己最早的发明创造。通过检索专利文献，我们发现，早在 1984 年特莱佛就提交过名为“轮椅的可调解扶手”的英国 GB2165192 号专利申请。

1991 年，特莱佛在收看一个关于艾滋病在非洲蔓延情况的电视节目时，一位预防艾滋病工作者的介绍引起了他的注意：控制艾滋病蔓延的一种有效方式是利用无线电广播，宣传普及预防艾滋病的知识；但在农村地区许多贫穷家庭连收音机用的电池都买不起。特莱佛想到，如果能设计一款使用类似闹钟发条来供电的收音机，就可以取代对于电池的需求。经过反复尝试，特莱佛终于制作出发条收音机的原型，上紧发条后能够持续运行 10 多分钟。特莱佛于 1992 年向英国递交了名为“电发生器”的 GB2262324 号专利申请，如图 22-1、图 22-2 所示。

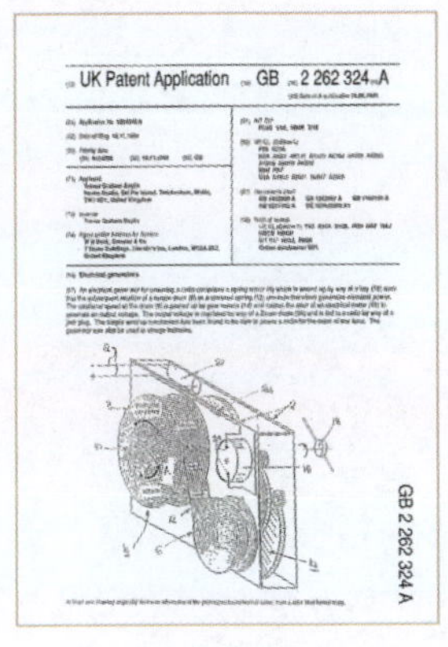

图 22-1

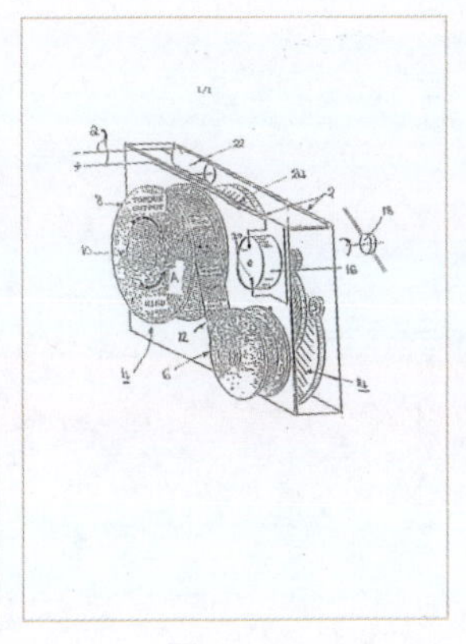

图 22-2

通过这篇专利文献，我们得以清楚地了解特莱佛的发明。特莱佛的专利申请保护的是发条收音机的供电装置，这也是发条收音机能够运行的核心。这种为无线电收音机提供电力的发电机包含一个弹簧马达，如图 22-2 所示。特莱佛将一根长长的弹簧片两头分别固定在一大一小两个滚筒上。弹簧片在松弛状态时缠绕在较小的滚筒上，在上紧发条时，弹簧片逐步缠绕到较大的滚筒上，上发条时获得的机械能转换成势能在发条内储存起来。当发条松开时，储存的能量通过齿轮传动装置和旋转电机，产生输出电压，再经齐纳二极管调节后，以插头的方式为收音机供电。

通过进一步检索，我们发现，特莱佛的这一项专利申请并没有获得授权。在对发明进行改进后，1995 年特莱佛再次在英国提交了专利申请，并于 1999 年获得英国 GB2304208 号专利权，如图 22-3、图 22-4、图 22-5 所示。在这项后续的发明中，特莱佛改进了发电机的齿轮传动系统和电路设计，使发电机的供电更可控制和更加稳定。

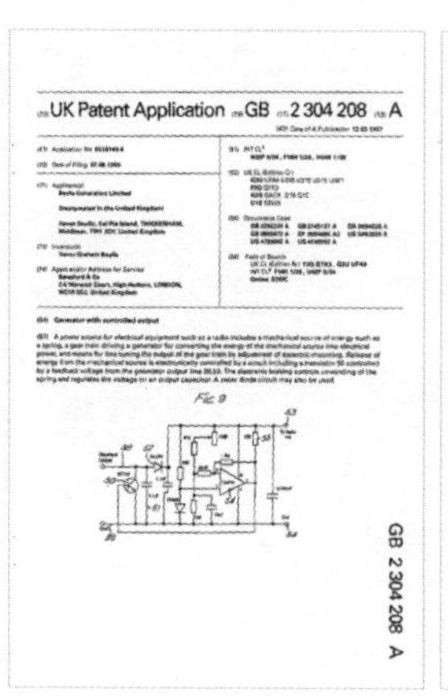

UK Patent Application GB 2 304 208 A

GB 2 304 208 A

图 22-3

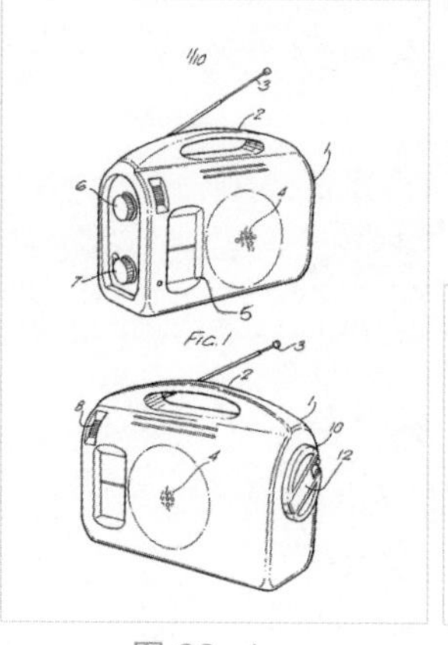

图 22-4

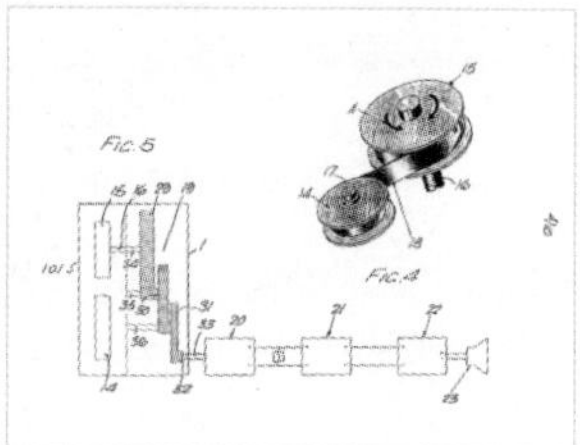

图 22-5

1994 年特莱佛成立了自由播放能源公司（Freeplay Energy），将他的发明投入市场。1995 年南非开普敦的工厂开始生产这种产品。产品上市为当地的人们带来了福音，预防艾滋病的广播宣传起到了明显效果。而且由于生产厂家主要雇佣残疾人作为工人，也解决了这部分人的就业和生计问题。

在改进发条收音机的技术方案的同时，特莱佛也在不断拓展它的市场。在 1995 年再次在英国提交专利申请后，他还将这一发明向美国、加拿大、德国、西班牙、澳大利亚、日本等多国提交了专利申请——这有助于特莱佛在这些国家的市场上获得独占权，使他的知识产权得到保护。

1997 年，针对西方市场的更为轻便的发条收音机上市，这一代产品仅需上发条 20 秒，就能够运行 1 个小时。这款产品上市后，成为陪伴时髦人士野营和其他露天活动的新宠，销售量很快超过 100 万台。

自由播放能源公司一直致力于研发其他手摇发电的绿色节能电子消费产品，包括手电筒、CD 机、移动电话，甚至笔记本电脑。仅需 24 毫瓦的电能、手动 1 分钟可带来 6 分钟光明的上发条电筒颇受市场欢迎。自由播放能源公司也为这些产品申请了专利。通过对比这些专利文献，可以看到它们的核心供电技术都继承自最初的发条收音机，如图 22-6、图 22-7、图 22-8 所示。例如，该公司于 1997 年在英国

申请的关于手摇发电手电筒的专利。

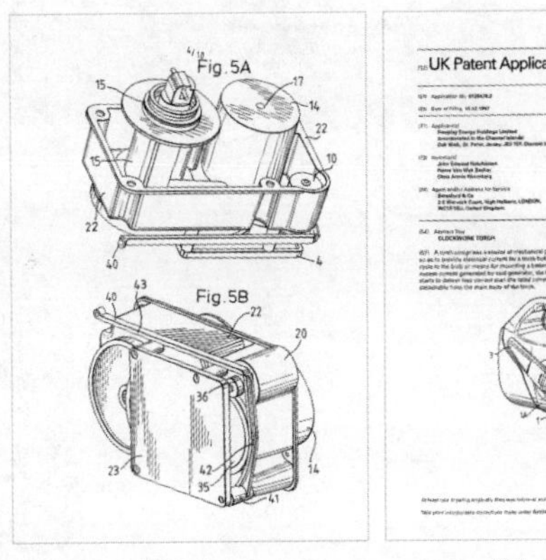

图 22-6

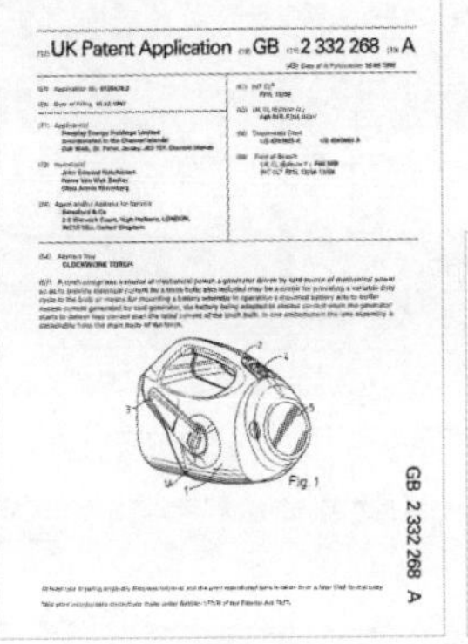
UK Patent Application GB 2 332 268 A

Fig.1

GB 2 332 268 A

图 22-7

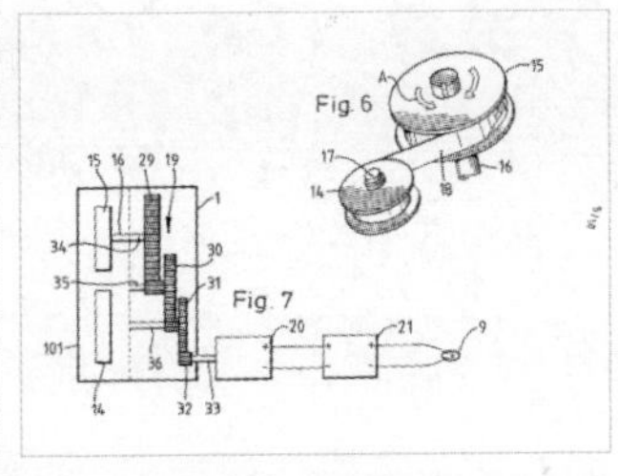

图 22-8

将英国 GB2332268 号专利附图（图 22-8）与英国 GB2304208 号专利附图（图 22-5）对比，两组附图几乎相同，说明手摇手电筒内部的供电马达延续了发条收音机的供电马达的设计。

由于发明发条收音机这一突出贡献，特莱佛获得了众多荣誉和嘉奖。他曾获得不列颠帝国勋章以及多所大学的荣誉学位，他的发条收音机曾获得英国广播公司颁发的设计奖（BBC Design Award）。作为一名发明家，由于对发明创造过程中的各种艰难深有体会，特莱佛以他的名字成立了基金会——特莱佛·贝利斯基金会（Trevor Baylis Foundation），为发明人、工程师们提供专业的帮助，鼓励和支持他们将创意灵感付诸实践，通过专利来保护自己的发明创造，以及将产品推向市场。特莱佛还非常关注中国在知识产权保护方面的发展。2010 年，英国一家电热水壶温控器生产商在中国诉两家中国企业侵犯其专利权，并获得胜诉。特莱佛对此专门致信中国驻英国大使，对中国法院的裁决表示了赞同和感谢，他赞扬中国对知识产权的保护，认为这将使外国企业对自己的知识产权在中国市场受到保护更有信心，同时这也为全球各国的知识产权保护作出了很好的示范。

1993. GB 2262324

——电发生器

UK Patent Application GB 2 262 324 A

(43) Date of A publication 16.06.1993

(22) Date of filing 19.11.1992

(71) Applicant
Trevor Graham Baylis
Haven Studio, Eel Pie Island, Twickenham, Middx, TW1 3DY, United Kingdom

(72) Inventor
Trevor Graham Baylis

(74) Agent and/or Address for Service
W H Beck, Greener & Co
7 Stone Buildings, Lincoln's Inn, London, WC2A 3SZ, United Kingdom

(54) Electrical generators

(57) An electrical generator for powering a radio comprises a spring motor (4) which is wound up by way of a key (18) such that the subsequent rotation of a torque drum (8) as a stressed spring (12) unwinds therefrom generates electrical power. The rotational speed of the drum (8) is geared up by gear means (14) and rotates the rotor of an electrical motor (16) to generate an output voltage. The output voltage is regulated by way of a Zener diode (24) and is fed to a radio by way of a jack plug. The simple wind up mechanism has been found to be able to power a radio for the order of one hour. The generator can also be used to charge batteries.

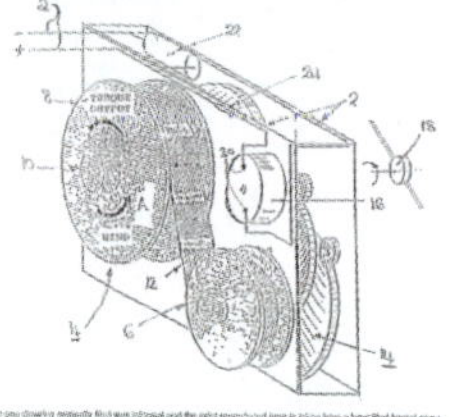

GB 2 262 324 A

1/3

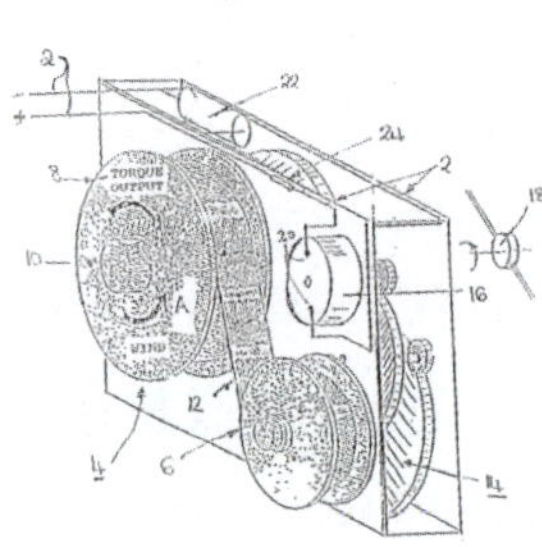

2262324

-1-

IMPROVEMENTS IN OR RELATING TO ELECTRICAL GENERATORS

The present invention relates to electrical generators, and to electrical devices powered by such generators.

In the Third World there are locations without mains electrical power. This means that any electrical devices, for example, such as radios, need to be battery operated. However, in such regions, batteries can be difficult to obtain, and more significantly, may be relatively expensive as compared to the resources of the people.

It is an object of the present invention to seek to make such electrical devices as may be available readily usable in such regions.

According to the present invention there is provided an electrical generator for powering an electrical device, said electrical generator comprising a spring motor having a metal band spring arranged to be stressed by being wound onto a rotatable torque drum and arranged to unwind from said drum to cause rotation thereof, an electrical motor having a rotor arranged to be rotated by the rotation of said torque drum and to thereby generate an electrical supply voltage at output terminals of said motor, gear means for coupling said torque drum to said rotor and arranged, upon the rotation of said drum, to rotate said rotor at a predetermined speed of rotation, and voltage regulator means electrically coupled to the output terminals of said motor such that the rotation of said torque drum causes a substantially constant voltage output.

An electrical generator of the invention may be used to power an electrical device, as a radio. As the power is generated mechanically from the spring motor no electrical

-2-

mains power is required for the operation of the device. Furthermore, the store of mechanical energy may be replenished by physical effort, that is, by rewinding the spring. There is no necessity to purchase batteries or other energy sources in order to keep the electrical device in working order.

Theoretically, the electrical device to be powered by the electrical generator may be any type of electrically powered device currently available. Preferably, however, the electrical device is portable and is one which is unlikely to require substantial electrical power levels. The electrical device may be, for example, a radio, a personal cassette player, a lamp, a games console, or a lap top computer.

In one embodiment, the electrical device comprises one or more batteries connected to be charged up by said electrical generator.

The spring motor may comprise a logarithmic spring or a spring conforming to Hookes' law. However, in a presently preferred embodiment, said spring motor has a metal band spring arranged to deliver a substantially constant force and thereby to cause said torque drum to rotate at a substantially constant speed of rotation.

For example, said metal band spring is a prestressed band of steel.

In a preferred embodiment, said spring motor comprises a freely rotatable, storage drum arranged such that the axes of said torque and storage drums extend substantially parallel, and wherein each end of the metal band spring is fastened to a respective one of the torque and storage drums such that the metal band spring may be wound from one

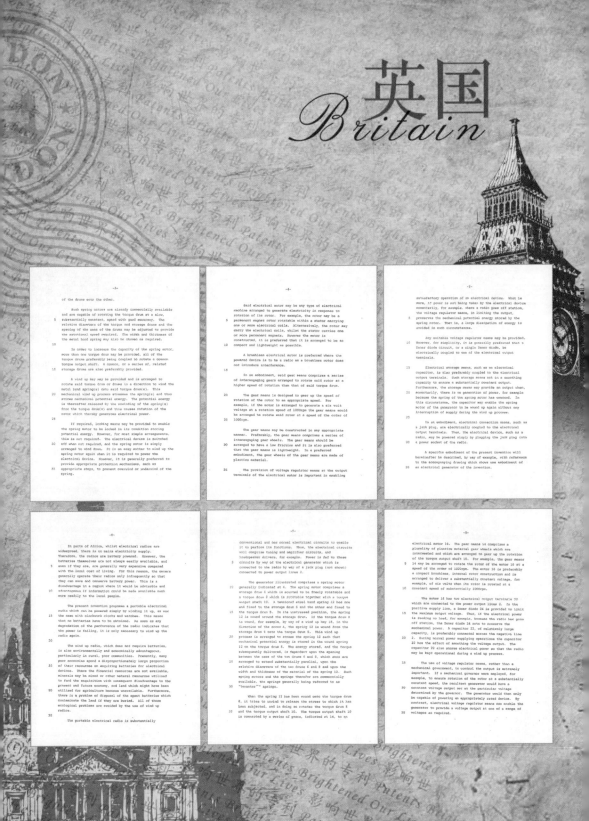

-3-

of the drums onto the other.

Such spring motors are already commercially available and are capable of rotating the torque drum at a slow, substantially constant, speed with good accuracy. The relative diameters of the torque and storage drums and the spacing of the axes of the drums may be adjusted to provide the rotational speed required. The width and thickness of the metal band spring may also be chosen as required.

In order to increase the capacity of the spring motor, more than one torque drum may be provided, all of the torque drums preferably being coupled to rotate a common torque output shaft. A common, or a series of, related storage drums are also preferably provided.

A wind up key may be provided and is arranged to rotate said torque drum or drums in a direction to wind the metal band spring(s) onto said torque drum(s). This mechanical wind up process stresses the spring(s) and thus stores mechanical potential energy. The potential energy is thereafter released by the unwinding of the spring(s) from the torque drum(s) and this causes rotation of the rotor which thereby generates electrical power.

If required, locking means may be provided to enable the spring motor to be locked in its condition storing potential energy. However, for most simple arrangements, this is not required. The electrical device is switched off when not required, and the spring motor is simply arranged to wind down. It is an easy matter to wind up the spring motor again when it is required to power the electrical device. However, it is generally preferred to provide appropriate protection mechanisms, such as appropriate stops, to prevent overwind or underwind of the spring.

-4-

Said electrical motor may be any type of electrical machine arranged to generate electricity in response to rotation of its rotor. For example, the rotor may be a permanent magnet rotor rotatable within a stator carrying one or more electrical coils. Alternatively, the rotor may carry the electrical coils, whilst the stator carries one or more permanent magnets. However the motor is constructed, it is preferred that it is arranged to be as compact and lightweight as possible.

A brushless electrical motor is preferred where the powered device is to be a radio as a brushless motor does not introduce interference.

In an embodiment, said gear means comprises a series of interengaging gears arranged to rotate said rotor at a higher speed of rotation than that of said torque drum.

The gear means is designed to gear up the speed of rotation of the rotor to an appropriate speed. For example, if the motor is arranged to generate a six volt voltage at a rotation speed of 1000rpm the gear means would be arranged to rotate said rotor at a speed of the order of 1000rpm.

The gear means may be constructed in any appropriate manner. Preferably, the gear means comprises a series of interengaging gear wheels. The gear means should be arranged to have a low friction and it is also preferred that the gear means is lightweight. In a preferred embodiment, the gear wheels of the gear means are made of plastics material.

The provision of voltage regulator means at the output terminals of the electrical motor is important in enabling

-5-

satisfactory operation of an electrical device. What is more, if power is not being taken by the electrical device momentarily, for example, where a radio goes off station, the voltage regulator means, in limiting the output, preserves the mechanical potential energy stored by the spring motor. That is, a large dissipation of energy is avoided in such circumstances.

Any suitable voltage regulator means may be provided. However, for simplicity, it is generally preferred that a Zener diode circuit, or a single Zener diode, be electrically coupled to one of the electrical output terminals.

Electrical storage means, such as an electrical capacitor, is also preferably coupled to the electrical output terminals. Such storage means act in a smoothing capacity to ensure a substantially constant output. Furthermore, the storage means may provide an output when, momentarily, there is no generation of power, for example because the spring of the spring motor has unwound. In this circumstance, the capacitor may enable the spring motor of the generator to be wound up again without any interruption of supply during the wind up process.

In an embodiment, electrical connection means, such as a jack plug, are electrically coupled to the electrical output terminals. Thus, the electrical device, such as a radio, may be powered simply by plugging the jack plug into a power socket of the radio.

A specific embodiment of the present invention will hereinafter be described, by way of example, with reference to the accompanying drawing which shows one embodiment of an electrical generator of the invention.

-6-

In parts of Africa, whilst electrical radios are widespread, there is no mains electricity supply. Therefore, the radios are battery powered. However, the batteries themselves are not always easily available, and even if they are, are generally very expensive compared with the local cost of living. For this reason, the owners generally operate their radios only infrequently so that they can save and conserve battery power. This is a disadvantage in a region where it would be advisable and advantageous if information could be made available much more readily to the local people.

The present invention proposes a portable electrical radio which can be powered simply by winding it up, as was the case with clockwork clocks and watches. This means that no batteries have to be obtained. As soon as any degradation of the performance of the radio indicates that the power is failing, it is only necessary to wind up the radio again.

The wind up radio, which does not require batteries, is also environmentally and economically advantageous, particularly in rural, poor communities. Presently, many poor economies spend a disproportionately large proportion of their resources on acquiring batteries for electrical devices. Where the financial resources are not available, minerals may be mined or other natural resources utilised to fund the acquisition with consequent disadvantage to the present and future economy, and land which might have been utilised for agriculture becomes unavailable. Furthermore, there is a problem of disposal of the spent batteries which contaminate the land if they are buried. All of these ecological problems are avoided by the use of wind up radios.

The portable electrical radio is substantially

-7-

conventional and has normal electrical circuits to enable it to perform its functions. Thus, the electrical circuits will comprise tuning and amplifier circuits, and loudspeaker drivers, for example. Power is fed to these circuits by way of the electrical generator which is connected to the radio by way of a jack plug (not shown) connected to power output lines 2.

The generator illustrated comprises a spring motor generally indicated at 4. The spring motor comprises a storage drum 6 which is mounted to be freely rotatable and a torque drum 8 which is rotatable together with a torque output shaft 10. A tensioned steel band spring 12 has one end fixed to the storage drum 6 and the other end fixed to the torque drum 8. In its unstressed position, the spring 12 is wound around the storage drum. If the torque drum 8 is wound, for example, by way of a wind up key 18, in the direction of the arrow A, the spring 12 is wound from the storage drum 6 onto the torque drum 8. This wind up process is arranged to stress the spring 12 such that mechanical potential energy is stored in the wound spring 12 on the torque drum 8. The energy stored, and the torque subsequently delivered, is dependent upon the spacing between the axes of the two drums 6 and 8, which axes are arranged to extend substantially parallel, upon the relative diameters of the two drums 6 and 8 and upon the width and thickness of the material of the spring 12. Such spring motors and the springs therefor are commercially available, the springs generally being referred to as "Tensator™" springs.

When the spring 12 has been wound onto the torque drum 8, it tries to unwind to release the stress to which it has been subjected, and in doing so rotates the torque drum 8 and the torque output shaft 10. The torque output shaft 10 is connected by a series of gears, indicated at 14, to an

-8-

electrical motor 16. The gear means 14 comprises a plurality of plastics material gear wheels which are intermeshed and which are arranged to gear up the rotation of the torque output shaft 10. For example, the gear means 14 may be arranged to rotate the rotor of the motor 16 at a speed of the order of 1000rpm. The motor 16 is preferably a compact brushless, internal rotor construction and is arranged to deliver a substantially constant voltage, for example, of six volts when its rotor is rotated at a constant speed of substantially 1000rpm.

The motor 16 has two electrical output terminals 20 which are connected to the power output lines 2. In the positive supply line, a Zener diode 24 is provided to limit the maximum output voltage. Thus, if the electrical power is feeding no load, for example, because the radio has gone off station, the Zener diode 24 acts to conserve the mechanical power. A capacitor 22, of relatively large capacity, is preferably connected across the negative line 2. During normal power supplying operations the capacitor 22 has the effect of smoothing the voltage output. The capacitor 22 also stores electrical power so that the radio may be kept operational during a wind up process.

The use of voltage regulator means, rather than a mechanical government, to control the output is extremely important. If a mechanical governor were employed, for example, to ensure rotation of the rotor at a substantially constant speed, the resultant generator would have a constant voltage output set at the particular voltage determined by the governor. The generator would then only be capable of powering an appropriately rated device. By contrast, electrical voltage regulator means can enable the generator to provide a voltage output at one of a range of voltages as required.

英国

Britain

-9-

In the embodiment illustrated, the voltage regulator means comprises the Zener diode 14. The output voltage may therefore be selected by selection of an appropriately rated Zener diode. In an alternative embodiment, not illustrated, a Zener diode circuit may be provided incorporating a number of differently rated Zener diodes and switch means operable to switch selected ones or combinations of the Zener diodes into the voltage output circuit. In this manner, the generator may be made comparable with, and thus able to power, a range of electrical devices.

The arrangement shown in the Figure is capable of providing power to operate a transistor radio for the order of one hour. It is then only necessary to re-wind the spring 12 onto the torque drum 8 to continue the radio operation.

Because the generator shown in the Figure is a stand alone device, employing the jack plug to make the electrical connection to the radio, it may also be used for other applications where electrical power is required.

In some poor countries, jack plugs and other connectors may not be available. Of course, the electrical connections may then be made by connecting the output lines 2 directly to the radio, and appropriately holding them in place.

The generator illustrated may alternatively be made sufficiently compact that it may be incorporated into the radio, for example in the space traditionally provided for batteries.

Of course, it will be appreciated that variations and modifications to the invention as described above may be

-10-

made. The mechanical energy for operating the radio may be generated and stored in any appropriate manner. Similarly, the electrical device powered by the stored mechanical energy may be a device other than a radio.

Whilst the invention has been developed specifically with the problems of Third World regions in mind, it also has other applications. For example, travellers may find it more convenient to carry mechanically powered electrical devices on their travels so that they are not let down if the battery runs out. Travellers will not then have to carry spare batteries or to try to find compatible replacements in a foreign country.

The invention is also particularly useful for emergency applications where an electrical device is not used frequently but must be available for use reliably. For example, if a battery powered radio and/or transmitter is provided in a liferaft or lifeboat it is currently necessary to check the batteries frequency to ensure power is available as and when required. A wind up device has the advantage that, once it has been wound up, it reliably provides power.

It will be appreciated that other modifications and improvements to the invention may be made within the scope of this application.

-11-

CLAIMS

1. An electrical generator for powering an electrical device, said electrical generator comprising a spring motor having a metal band spring arranged to be stressed by being wound onto a rotatable torque drum and arranged to unwind from said drum to cause rotation thereof, an electrical motor having a rotor arranged to be rotated by the rotation of said torque drum and to thereby generate an electrical supply voltage at output terminals of said motor, gear means for coupling said torque drum to said rotor and arranged, upon the rotation of said drum, to rotate said rotor at a predetermined speed of rotation, and voltage regulator means electrically coupled to the output terminals of said motor such that the rotation of said torque drum causes a substantially constant voltage output.

2. An electrical generator as claimed in Claim 1, wherein said spring motor has a metal band spring arranged to deliver a substantially constant force and thereby to cause said torque drum to rotate at a substantially constant speed of rotation.

3. An electrical generator as claimed in Claim 1 or Claim 2, wherein said metal band spring is a prestressed band of steel.

4. An electrical generator as claimed in any preceding claim, wherein said spring motor comprises a freely rotatable, storage drum arranged such that the axes of said torque and storage drums extend substantially parallel, and wherein each end of the metal band spring is fastened to a respective one of the torque and storage drums such that the metal band spring may be wound from one of the drums onto the other.

-12-

5. An electrical generator as claimed in any preceding claim, further comprising a wind up key arranged to rotate said torque drum in a direction to wind said metal band spring onto said torque drum.

6. An electrical generator as claimed in any preceding claim, wherein said electrical motor is a compact, lightweight, brushless, electrical motor.

7. An electrical generator as claimed in any preceding claim, wherein said gear means comprises a series of interengaging gears arranged to rotate said rotor at a higher speed of rotation than that of said torque drum.

8. An electrical generator as claimed in Claim 7, wherein said gear means is arranged to rotate said rotor at a speed of the order of 1000rpm.

9. An electrical generator as claimed in Claim 7 or 8, wherein said gear means comprise a series of interengaging gear wheels.

10. An electrical generator as claimed in any of Claims 7 to 9, wherein said gear means are made of a plastics material.

11. An electrical generator as claimed in any preceding claim, wherein said voltage regulator means comprises a Zener diode electrically coupled to one of said electrical output terminals.

12. An electrical generator as claimed in any preceding claim, further comprising electrical storage means coupled to said electrical output terminals.

13. An electrical generator as claimed in Claim 12,

-13-

wherein said electrical storage means comprises a capacitor.

14. An electrical generator as claimed in any preceding claim, further comprising electrical connection means electrically coupled to said electrical output terminals.

15. An electrical device comprising electrical circuit means for performing the functions of said electrical device, and an electrical generator as claimed in any preceding claim coupled to supply electrical power to said electrical circuit.

16. An electrical generator substantially as hereinbefore described with reference to the accompanying drawing.

17. An electrical device substantially as hereinbefore described with reference to the accompanying drawing.

14

Patents Act 1977
Examiner's report to the Comptroller under Section 17 (The Search Report)

Application number: GB 9224246.5

Relevant Technical fields

(i) UK Cl (Edition L) T2B (B3A, B22B); H2A (AACB, UBCU, RAF, UAJ)

(ii) Int Cl (Edition 5) H02K, H02J

Search Examiner: D DODD

Databases (see over)

(i) UK Patent Office

(ii) ONLINE DATABASES: WPI

Date of Search: 25 FEBRUARY 1993

Documents considered relevant following a search in respect of claims 1

Category (see over)	Identity of document and relevant passages		Relevant to claim(s)
Y	GB 1432960	(SOLID) - Note Figure	1 at least
Y	GB 1262847	(SOLID) - Note eg Figure 5	1 at least
Y	GB 0217452	(ROTTURA) - Note Figure 3	1 at least
Y	DE 3442842 A1	(WIDGMANN) - Note Figure 1	1 at least
Y	GB 1186165	(GEC) - Note example of voltage regulator 11 in Figure 3	1 at least

SF2(p) NF - doc99\fil000090

Category	Identity of document and relevant passages	Relevant to claim(s)

-15-

Categories of documents

X: Document indicating lack of novelty or of inventive step.

Y: Document indicating lack of inventive step if combined with one or more other documents of the same category.

A: Document indicating technological background and/or state of the art.

P: Document published on or after the declared priority date but before the filing date of the present application.

E: Patent document published on or after, but with priority date earlier than, the filing date of the present application.

&: Member of the same patent family, corresponding document.

Databases: The UK Patent Office database comprises classified collections of GB, EP, WO and US patent specifications as outlined periodically in the Official Journal (Patents). The on-line databases considered for search are also listed periodically in the Official Journal (Patents).

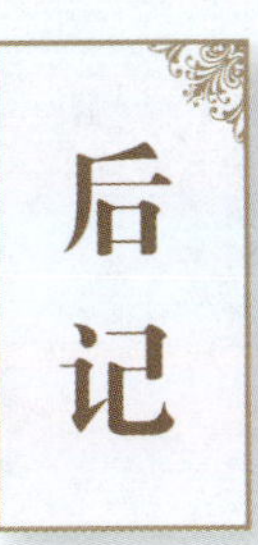

后记

创新是人类社会进步的“原动力”。回顾人类社会发展历程，从筷子、车轮的创意到“四大发明”，从衣食住行的方方面面到关系国计民生的尖端科技，人类绵延数千年的辉煌文明是一代又一代创新者智慧与创造的结晶。电灯不是爱迪生凭空想象出来的，地球重力的发现也不是牛顿一个人的贡献，“小众”的成功是“站在巨人的肩膀上”获得的，影响世界的专利离不开“大众”的不懈尝试与努力。随着现代专利制度的诞生，凝聚着人类智慧与心血的发明创造和构想被完整系统地记录在专利文献中，成为大众创业、万众创新不可缺少的“工具箱”。专利文献不仅记载了发明创造的完整技术方案，而且揭示了发明人的创新历程和技术发展脉络。在创新活动中通过利用专利文献可以帮助创新者开阔思路、激发灵感，在他人智慧成果的基础上寻找最佳创新方案。

本书甄选了22项发明专利，向广大读者呈现了为人类社会做出重大贡献的发明家们是如何运用专利制度，从“草根”创业者努力走向成功的故事。虽然其中很多发明家的故事早已是家喻户晓，但编者从专利文献的视角，梳理他们在创新和申请专利保护过程中的艰辛历程，解读他们的专利技术对后续创新者的影响，并通过突出呈现诸如专利权纠纷、企业专利布局及专利权成功商业转化等具体案例，帮助读者了解专利制度对创新活动的促进和保护作用，传播创业创新文化，激发全民族的创业精神和创新热情，营造大众创业、万众创新的良好氛围。

编辑组成员为：朱跃，王亚玲，王英丽，卢世超，盖爽，贾丹明，

隆捷，李鸿斌，黄迎燕，杨策，郑宁，吴泉洲，高会霞，吴小松，刘乃莛，邹春青，刘小青。王亚玲编写第三篇、第十六篇及第十七篇；黄迎燕编写第二篇、第十八篇；隆捷编写第八篇、第十三篇、第二十二篇；李鸿斌编写第六篇、第十篇；杨策编写第九篇、第十九篇；卢世超编写第一篇、第十五篇；郑宁编写第四篇、第七篇；盖爽编写第五篇、第二十篇；王英丽编写第十一篇、第十二篇；贾丹明编写第十四篇、第二十一篇。吴泉洲、高会霞、吴小松、刘乃莛、邹春青、刘小青协助完成专利文献检索、素材搜集整理工作。朱跃、王亚玲、隆捷、王英丽对全书进行审校与统稿。

本书甄选的22项发明故事是在2009年国家知识产权局专利局专利文献部制作的《影响世界的专利》宣传画基础上，基于专利文献资源，广泛结合互联网信息资源汇编而成。其中，参考了国家知识产权局网站《影响世界的专利》栏目（网址http://www.sipo.gov.cn/ztzl/ywzt/yxsjdip/）中的多篇文章，在此我们向诸位作者表示感谢。由于我们学识水平有限，虽然经过检索、阅读大量文献资料，并多次反复讨论、推敲、修改，但依然存在不足，恳请批评指正，并提出宝贵意见与建议。

编辑组

2015年10月

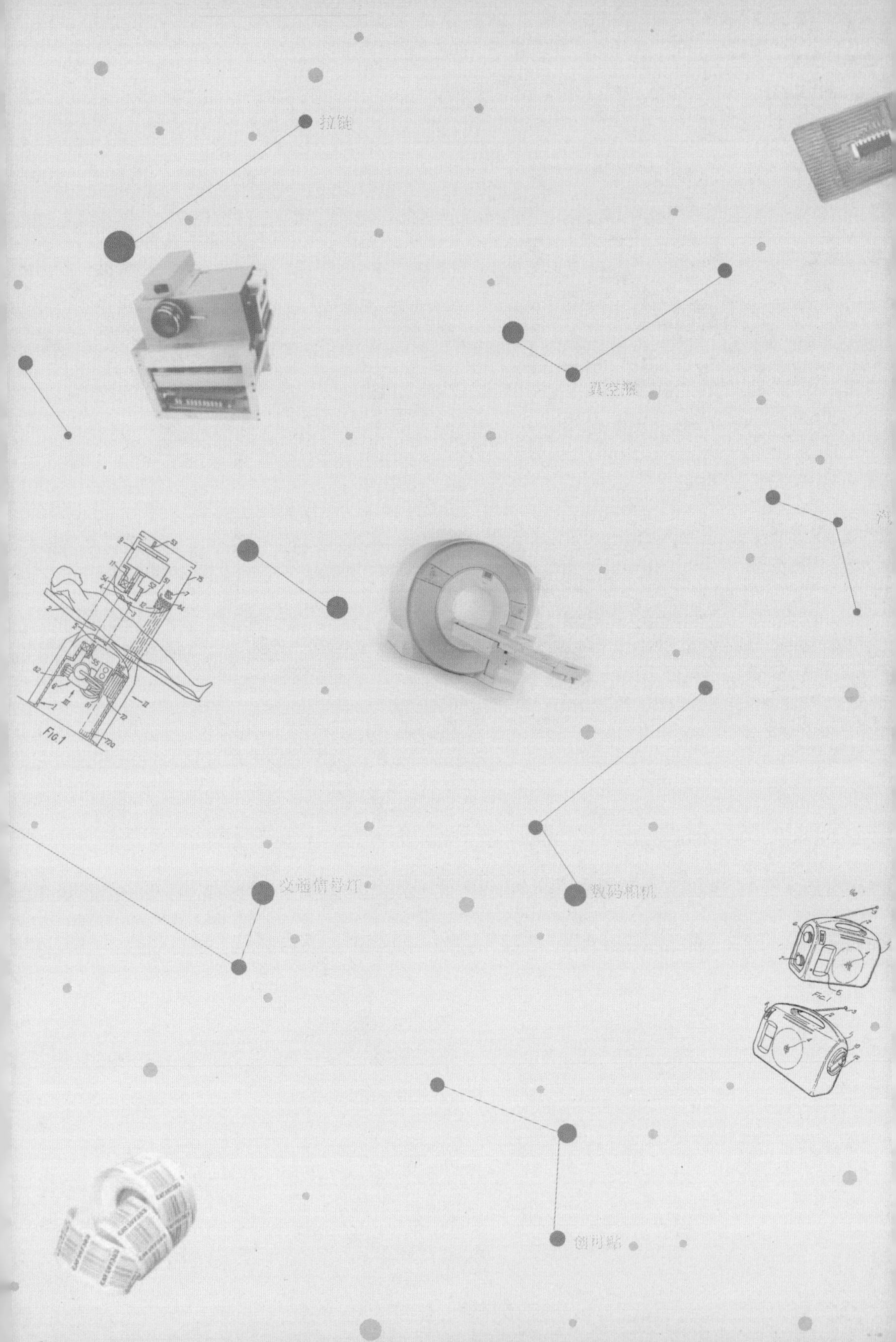
拉链
真空瓶
交通信号灯
数码相机
创可贴